Keeping Things in Proportion
Reasoning with Ratios

STUDENT BOOK

TERC

Mary Jane Schmitt, Myriam Steinback,
Donna Curry, Tricia Donovan, Martha Merson

McGraw Hill Education

Bothell, WA • Chicago, IL • Columbus, OH • New York, NY

TERC
2067 Massachusetts Avenue
Cambridge, Massachusetts 02140

T E R C

EMPower Research and Development Team
Principal Investigator: Myriam Steinback
Co-Principal Investigator: Mary Jane Schmitt
Research Associate: Martha Merson
Curriculum Developers: Donna Curry, Tricia Donovan

Technical Team
Production and Design Coordinator: Valerie Martin
Project Assistant and Graphic Designer: Juania Ashley
Copyeditor: Jill Pellarin

Evaluation Team
Brett Consulting Group:
    Belle Brett
    Marilyn Matzko

EMPower™ was developed at TERC in Cambridge, Massachusetts. This material is based upon work supported by the National Science Foundation under award number ESI-9911410 and by the Education Research Collaborative at TERC. Any opinions, findings, and conclusions or recommendations expressed in this publication are those of the authors and do not necessarily reflect the views of the National Science Foundation.

TERC is a not-for-profit education research and development organization dedicated to improving mathematics, science, and technology teaching and learning.

All other registered trademarks and trademarks in this book are the property of their respective holders.

http://empower.terc.edu

Printed in the United States of America
4 5 6 7 8 9 QVS 19 18 17 16 15

ISBN 978-0-07662-093-7
MHID 0-07-662093-X

Contents

Contents

Introduction

Welcome to EMPower

Students using the EMPower books often find that EMPower's approach to mathematics is different from the approach found in other math books. For some students, it is new to talk about mathematics and to work on math in pairs or groups. The math in the EMPower books will help you connect the math you use in everyday life to the math you learn in your courses.

We asked some students what they thought about EMPower's approach. We thought we would share some of their thoughts with you to help you know what to expect.

> *"It's more hands-on."*
>
> *"More interesting."*
>
> *"I use it in my life."*
>
> *"We learn to work as a team."*
>
> *"Our answers come from each other… [then] we work it out ourselves."*
>
> *"Real-life examples like shopping and money are good."*
>
> *"The lessons are interesting."*
>
> *"I can help my children with their homework."*
>
> *"It makes my brain work."*
>
> *"Math is fun."*

EMPower's goal is to make you think and to give you puzzles you will want to solve. Work hard. Work smart. Think deeply. Ask why.

Using This Book

This book is organized by lessons. Each lesson has the same format.

- The first page explains the lesson and states the purpose of the activities. Look for a question to keep in mind as you work.

- The activity pages come next. You will work on the activities in class, sometimes with a partner or in a group.

- Look for shaded boxes with additional information and ideas to help you get started if you become stuck.

- Practice pages follow the activities. These practices will make sense to you after you have done the activities. The three types of practice pages are

 Practice: Provides another chance to see the math from the activity and to use new skills.

 Extension: Presents a challenge with a more difficult problem or a new but related math idea.

 Test Practice: Asks a number of multiple-choice questions and one open-ended question.

In the *Appendices* at the end of the book, there is space for you to keep track of what you have learned and to record your thoughts about how you can use the information.

- Use notes, definitions, and drawings to help you remember new words in *Vocabulary*, pages 139–140.

- Answer the *Reflections* questions after each lesson, pages 141–144.

Tips for Success

Where do I begin?

Many people do not know where to begin when they look at their math assignments. If this happens to you, first try to organize your information. Read the problem. Start a drawing to show the situation.

This unit is about comparing amounts as rates and ratios.

Ask yourself:

What is the relationship between the amounts? How can I use that relationship to predict for other situations?

Another part of getting organized is figuring out what skills are required.

Ask yourself:

What do I already know? What do I need to find out?

Write down what you already know.

I cannot do it. It seems too hard.

Make a drawing or diagram.

Make the numbers smaller or friendlier. Try to solve the same problem with less complicated numbers to see the pattern.

Ask yourself:

Have I ever seen something like this before? What did I do then?

You can always look back at another lesson for ideas.

Am I done?

Don't walk away yet. Check your answers to make sure they make sense.

Ask yourself:

Did I answer the question?

Does the answer seem reasonable? Do the conclusions I am drawing seem logical?

Check your math with a calculator. Ask others whether your work makes sense to them.

Practice

Complete as many of the practice pages as you can to sharpen your skills.

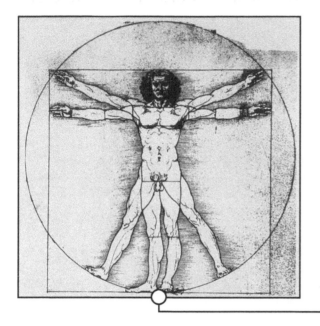

Opening the Unit: Comparing and Predicting

The mathematics people use in everyday life often has to do with making **comparisons** that help us answer questions such as: What's the better buy? or Which amount is larger? People also use mathematics to **predict**. They use evidence they already have to calculate for larger or smaller amounts.

In this session, you will use the strategies you already know to compare and predict. You may be able to do the problems in your head if the numbers are friendly. Otherwise, you may want to use a calculator or some paper-and-pencil method you know. By the end of the unit, you will have several more methods to choose from.

Activity 1: What Would It Take?

Imagine what it would take to fulfill your dreams. How much income **per** year would you say you and your family would need to fulfill all of your dreams?

1. Describe some of your dreams and the amount of money it might take to fulfill them. Include in your description your expected annual salary. Then be prepared to work with a partner to discuss your dream job.

2. Now, with a partner, take turns interviewing each other about the income you would need for your dreams. Use the interview questions on the next page and record your partner's answers. Your partner will record your answers on his or her page.

Interview Questions

Ask your partner to read his or her dream job description. Then ask the following questions:

1. What is the annual salary you would need to fulfill your dreams?

2. If you were paid the annual salary needed for your dreams, how much would you make monthly? How did you figure that out?

3. How much would you make weekly? How did you figure that out?

4. How does your dream salary compare with the salary for the job you have now (or one you have had in the past)?

5. What portion of your dream salary do you think would have to go to taxes?

6. Ask a question of your own, and write you partner's answer here.

Comparing Your Dreams

When you have each completed the interviews, use the information to make a poster with at least three statements that compare your dream salaries. Use fractions, decimals, percents, or differences to show how your dream salaries compare.

Activity 2: Initial Assessment: Comparing and Predicting Tasks

Your teacher will show you some problems and ask you to check off how you feel about your ability to solve them. In each case, check off one of the following:

___ Can do   ____ Don't know how   ____ Not sure

Activity 3: Mind Map

Make a Mind Map using words, numbers, pictures, or ideas that come to mind when you think of *rate*, *ratio*, and *proportion*.

RATE, RATIO, PROPORTION

Keeping Things In Proportion: Reasoning With Ratios Unit Goals

- Use a variety of strategies to examine the relationship within a ratio as well as the relationship between ratios.

- Determine whether ratios are equal.

- Solve for the missing value when three of the four amounts are known.

- Determine whether a situation is proportional.

My Own Goals

A Close Look at Supermarket Ads

Where do you see ratios used in supermarkets?

Rates and ratios show up in many everyday situations. Supermarkets use **ratios** to set prices, for example, 1 for $2.50, 2 for $5.00, 10 for $25.00, and so on.

This lesson is about working with **rates** and ratios in a variety of consumer situations. You will figure out patterns for equal ratios using pictures and rules so that you can clearly see what makes ratios equal.

You will determine when buying in large quantities results in saving money and when it does not.

Activity 1: Different Numbers—Same Ratio

Working with a partner, look through some supermarket flyers. Find and cut out two ads that use ratios.

Tape the two ads in the spaces provided below.

Write the original ratio in **fraction form**.

Create at least three other ratios that keep the same balance between the amount of food (or number of items) and the price.

Use a drawing to show that your ratios are all equal.

1. Place your first ad here.

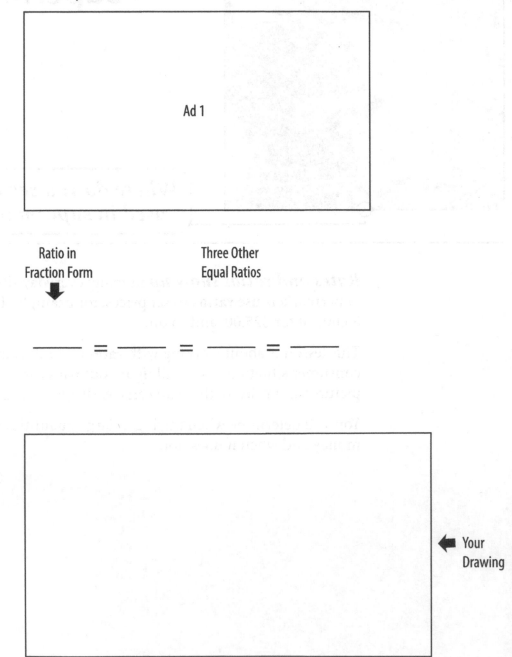

2. Place your second ad here.

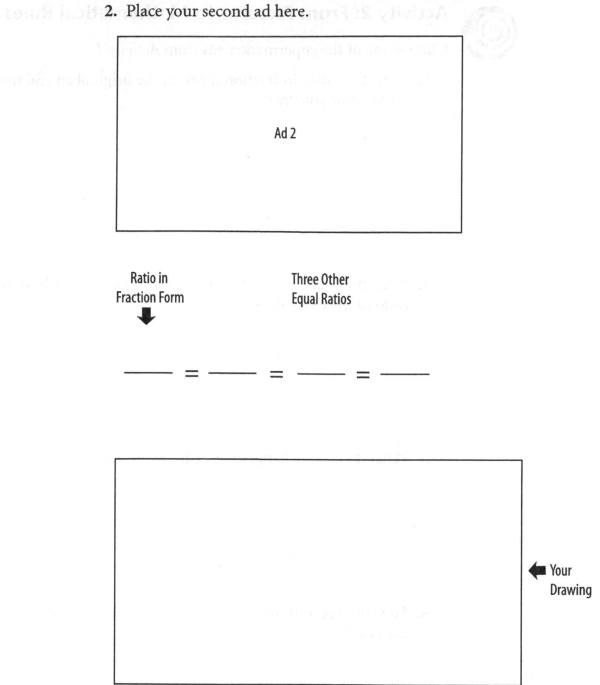

Ad 2

Ratio in
Fraction Form

Three Other
Equal Ratios

———— = ———— = ———— = ————

Your
Drawing

Activity 2: From Pictures to Mathematical Rules

Choose one of the supermarket ads from *Activity 1*.

1. Write the ratios in fraction form for the original ad and the other three ratios you created.

2. What mathematical operation might you use to get from the first ratio to the new ratios?

3. What do you see happening with the numbers?

4. How does your drawing in *Activity 1* match what you saw in the numbers?

Activity 3: Buying in Quantity

Some people love to go to stores where they can buy large quantities in bulk, but not everyone thinks this is a great idea.

1. Consider the pros and cons of buying in bulk.

    A. The pros: What are some good reasons for buying in bulk?

    B. The cons: What are some reasons for *not* buying in bulk?

2. Now take an ad from the flyers you used in *Activity 1* and, with a partner, create a new set of ads.

    Steps:

    *   Select an ad from the flyer you used in *Activity 1*. Write out the original ad.

    *   Create a new ad with the same ratio so that the deal is exactly the same whether you buy in large amounts or not.

    *   Create a new ad in which it is cheaper to buy in bulk than the offer in the original ad.

    *   Create a new ad in which it is more expensive to buy in bulk than the offer in the original ad.

    A. Original ad:

B. The same deal when buying in bulk:

C. A better deal when buying in bulk:

D. A worse deal when buying in bulk:

3. Put each of your ratios on a different index card. Then exchange your four cards with the cards that another pair of students made. Look at the other pair's cards and try to figure out which of the cards represents the ratio for the original ad and which show the same deal, a better deal, and a worse deal when buying in bulk.

   Mark an "A" on the card that you think shows the original ad.

   Mark a "B" on the card that shows the same deal when buying in bulk.

   Mark a "C" on the card that shows a better deal when buying in bulk.

   Mark a "D" on the card that shows a worse deal when buying in bulk.

   Compare your answers with those of the other student pair to see whether you agree. If there is disagreement, use diagrams or the **property of equal fractions** to show who is correct.

Practice: Balancing Ratios

Look at the sale prices from flyers below. Write another ratio that keeps the same balance. Use drawings to help you.

Example:

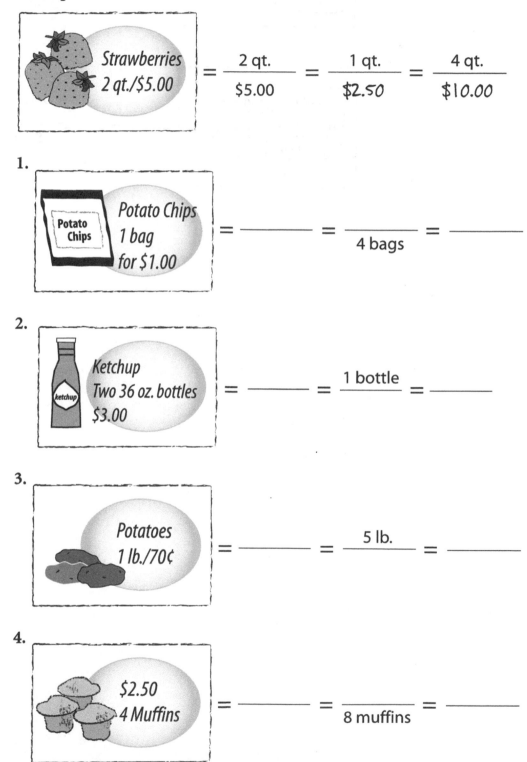

$$\frac{\text{Strawberries}}{2\ qt./\$5.00} = \frac{2\ qt.}{\$5.00} = \frac{1\ qt.}{\$2.50} = \frac{4\ qt.}{\$10.00}$$

1.

Potato Chips 1 bag for $1.00

$$= \frac{\quad}{\quad} = \frac{\quad}{4\ bags} = \frac{\quad}{\quad}$$

2.

Ketchup Two 36 oz. bottles $3.00

$$= \frac{\quad}{\quad} = \frac{1\ bottle}{\quad} = \frac{\quad}{\quad}$$

3.

Potatoes 1 lb./70¢

$$= \frac{\quad}{\quad} = \frac{5\ lb.}{\quad} = \frac{\quad}{\quad}$$

4.

$2.50 4 Muffins

$$= \frac{\quad}{\quad} = \frac{\quad}{8\ muffins} = \frac{\quad}{\quad}$$

5.

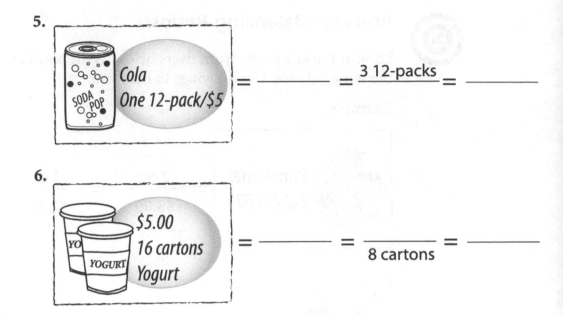

$$= \underline{} = \frac{3\ 12\text{-packs}}{} = \underline{}$$

6.

$$= \underline{} = \frac{}{8\ \text{cartons}} = \underline{}$$

Practice: Using the Property of Equal Fractions

In mathematics, a **proportion** is a statement that two ratios are equal.

For each proportion below describe what happened to each of the numbers in order to keep the ratio the same.

Example:

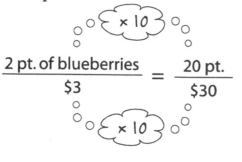

$$\frac{2 \text{ pt. of blueberries}}{\$3} = \frac{20 \text{ pt.}}{\$30}$$

Multiply the number of pints by 10 and mutiply the cost by 10.

1.

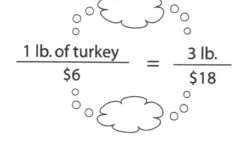

$$\frac{1 \text{ lb. of turkey}}{\$6} = \frac{3 \text{ lb.}}{\$18}$$

2.

$$\frac{2 \text{ cartons of yogurt}}{\$1} = \frac{30 \text{ cartons}}{\$15}$$

3.

$$\frac{2 \text{ bags of cookies}}{\$5} = \frac{1 \text{ bag}}{\$2.50}$$

4.

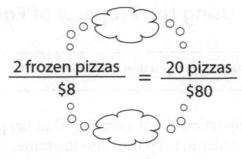

$$\frac{2 \text{ frozen pizzas}}{\$8} = \frac{20 \text{ pizzas}}{\$80}$$

5. Choose one of the situations in Problems 1–4. Show with pictures that the ratios are equal.

Practice: Is Bigger Cheaper?

Is It Worth Buying the Larger Size?

Many people think that buying the larger amount or bigger size is always a better deal than buying the lesser amount or the smaller size. But is it always so?

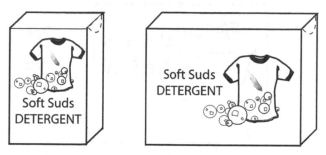

1. An ad shows a box of detergent selling for twice as much as a box with half as much detergent. Would you consider this a bargain? Explain your reasoning by drawing or using numbers.

2. Another ad shows a box of detergent that costs twice as much as a smaller box. The ad says the big box is a better deal than the smaller box. What would you predict is true about the larger box? (Check the answer that you think is the best prediction.)

    _____ It has more than twice the amount of detergent of the smaller box.

    _____ It has exactly twice the amount of detergent of the smaller box.

    _____ It has less than twice the amount of detergent of the smaller box.

Explain your reasoning with a drawing or numbers.

Practice: Working with Larger Numbers

For each problem, write some ratios that keep the balance of price to amount of food the same.

Example:

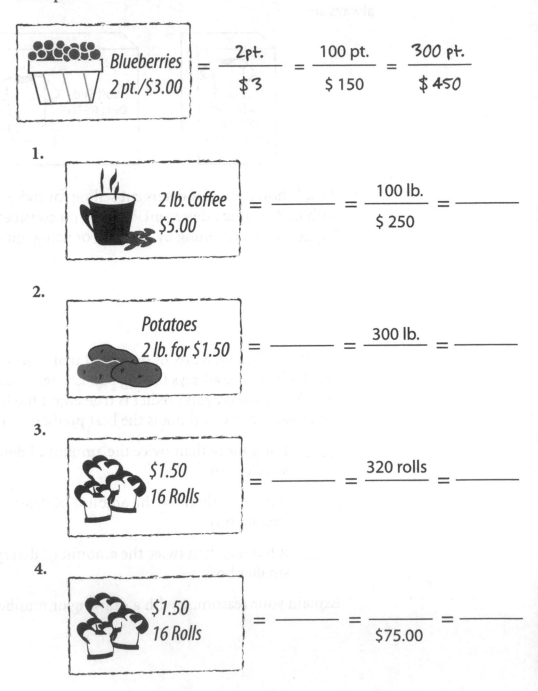

Blueberries 2 pt./$3.00 $= \dfrac{2 \text{ pt.}}{\$3} = \dfrac{100 \text{ pt.}}{\$150} = \dfrac{300 \text{ pt.}}{\$450}$

1.

2 lb. Coffee $5.00 $= \underline{\hspace{2cm}} = \dfrac{100 \text{ lb.}}{\$250} = \underline{\hspace{2cm}}$

2.

Potatoes 2 lb. for $1.50 $= \underline{\hspace{2cm}} = \dfrac{300 \text{ lb.}}{\hspace{1cm}} = \underline{\hspace{2cm}}$

3.

$1.50 16 Rolls $= \underline{\hspace{2cm}} = \dfrac{320 \text{ rolls}}{\hspace{1cm}} = \underline{\hspace{2cm}}$

4.

$1.50 16 Rolls $= \underline{\hspace{2cm}} = \dfrac{\hspace{1cm}}{\$75.00} = \underline{\hspace{2cm}}$

Practice: Mental Math

Create an equal ratio following the directions given. Try to do the multiplication and division in your head.

Example:

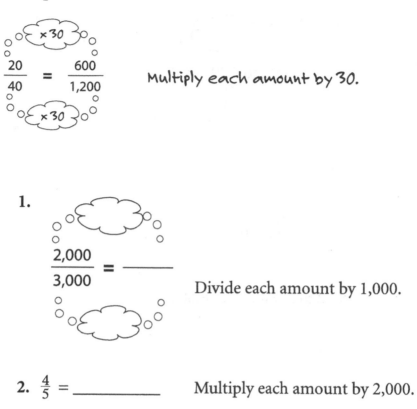

Multiply each amount by 30.

1.

$$\frac{2,000}{3,000} = \underline{\qquad}$$

Divide each amount by 1,000.

2. $\frac{4}{5} = \underline{\qquad\qquad}$ Multiply each amount by 2,000.

Write a rule for each of the following sets of ratios.

3.

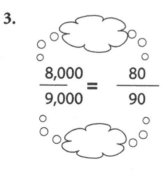

$$\frac{8,000}{9,000} = \frac{80}{90}$$

4.

10 to 30   =   2 to 6

5. $\dfrac{1}{4} = \dfrac{60}{240}$

6.

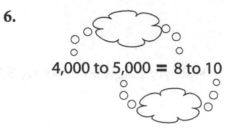

4,000 to 5,000 = 8 to 10

7. $\dfrac{100}{1,000} = \dfrac{1}{10}$

8. 2 to 3 = 4,000 to 6,000

Extension: Ratios in My Life

1. Think about all the ways that you see ratios in your daily life. List some of them in the columns below.

At Home	In the Community	At Work

2. Look at the ratios you listed. Do any of the ratios raise questions for you? Why?

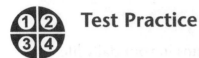
1. Soda is on sale at two 12-packs for $4. How much would 48 cans of soda cost at this rate?

   (1) $1.92

   (2) $2.00

   (3) $4.00

   (4) $4.80

   (5) $8.00

2. Fresh-baked onion rolls sell at 12 for $3. At this rate, how much would six cost?

   (1) $1.00

   (2) $1.50

   (3) $2.00

   (4) $2.50

   (5) $4.50

3. Lunchtime snacks cost $3 for four packages. Ramon wants to buy six packages so he has enough to last for the entire week. How much does he have to pay for them?

   (1) $4.00

   (2) $4.50

   (3) $5.00

   (4) $6.00

   (5) $6.50

4. Which deal is not the same as the others?

   (1) $4.50 for 10 lemons

   (2) Lemons: $6 per dozen

   (3) 5 lemons/$2.50

   (4) Lemons: 50 cents ea.

   (5) Two lemons for $1

5. Which is the best deal?

(1) Shop 'n Save Peaches
$1.29/lb.

(2) Shop 'n Save Peaches
2 lb. bag
only $2.50

(3) Shop 'n Save Peaches
Special
10 lb./$8.00

(4) Shop 'n Save Peaches
Buy 2 lb. @ $2
Get one lb. free

(5) Shop 'n Save Peaches
Supersaver
3 lbs. for $2.79

6. Two 6-packs of ice cream sandwiches cost $5. At this rate, how much would 60 6-packs cost?

It's a Lot of Work!

> *How does taking a sample help in predicting?*

We often take a sample—a portion or piece—to understand something about the bigger picture. For example, to find out what percent of American households have dogs, you could poll a small number of people and then use proportional reasoning to estimate how many dogs there are in the total number of U.S. households. In this lesson, you will conduct a **sample** of work to predict. Use the idea of keeping ratios equal to predict for larger amounts.

Activity: How Long Will It Take?

You offered to help at a community event and said you were willing to volunteer about four or five hours of your time. The coordinator took you up on your offer and assigned you the following jobs:

- Stuffing 1,000 flyers

- Peeling 50 pounds of potatoes for a huge potato salad

- Rolling 10,000 pennies from the penny-toss

You think these jobs are going to take a lot more time than the four or five hours you offered, but you would like to get a better idea of exactly how much time these jobs will take.

To get a good estimate of how much time these jobs require, you conduct a sample of each activity and use the property of equal fractions to figure out the time needed for the total job. How long do you estimate the assignment will take?

Work with your team to answer the questions for the activity at each station.

- Take a sample of how long it takes to do a part of the job.

- Use the sample rate to predict the time needed to complete the whole job. Show your reasoning.

- Be sure that everyone on your team can explain how the prediction was made. (This might require more than one strategy to explain the team's reasoning.)

Station 1. Stuffing Envelopes

How long would it take somebody to fold these flyers and stuff them into 1,000 envelopes? (Someone else will close, address, and stamp the envelopes.)

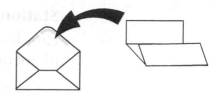

1. The sample we did and what we found out:

2. The method we used for figuring out how much time it would take to finish the job:

3. Our final estimate: _____

4. How we checked that the ratios are equal:

Station 2. Potato Salad

How long would it take somone to peel 50 pounds of potatoes?

1. The sample we did:

2. The method we used for figuring out how much time it would take to finish the job:

3. Our final estimate: _____

4. How we checked that the ratios are equal:

EMPower™

Station 3. Rolling Pennies

How long would it take somebody to roll 10,000 pennies?

1. The sample we did:

2. The method we used for figuring out how much time it would take to finish the job:

3. Our final estimate: _____

4. How we checked that the ratios are equal:

Summary

1.

Tasks	Our Time Estimate
Folding and stuffing 1,000 envelopes	
Peeling 50 lbs. of potatoes	
Rolling 10,000 pennies	
Total Time	

2. Compare your time estimates with the time estimates from other teams.

What would you say is a good estimate of time? Why?

Practice: Time to Be Proportional

Time is proportional. There are 60 seconds in one minute, every 60 minutes is one hour, and every 24 hours is one day. See how quickly you can figure out the missing numbers in these time-related ratios. Do the math in your head, using what you know about the rule of equal fractions.

1. $\dfrac{24 \text{ hours}}{1 \text{ day}} = \dfrac{\underline{\hspace{2cm}}}{3 \text{ days}}$

2. $\dfrac{60 \text{ minutes}}{1 \text{ hours}} = \dfrac{180 \text{ minutes}}{\underline{\hspace{2cm}}}$

3. $\dfrac{48 \text{ hours}}{\text{days}} = \dfrac{72 \text{ hours}}{\text{days}}$

4. $\dfrac{120 \text{ seconds}}{\text{minutes}} = \dfrac{3,600 \text{ seconds}}{\text{minutes}}$

5. $\dfrac{2 \text{ minutes}}{\text{seconds}} = \dfrac{4 \text{ minutes}}{\underline{\hspace{2cm}}}$

6. $\dfrac{12 \text{ hours}}{\text{days}} = \dfrac{\underline{\hspace{2cm}}}{2 \text{ days}}$

7. Choose one of the proportions above and use a diagram to check your answer.

Practice: Wasting Water

How much water does a dripping faucet waste in a year? That is an important question as we become more and more concerned about limited resources such as clean water.

Try an experiment: Take three samples of faucets dripping in your own home. Predict how much water would drip from those three faucets together in a year. If you do not have dripping faucets, turn your faucet on to drip a bit for the purposes of the experiment.

1. How I chose my samples:

2. Drip samples:

   Sample 1: _____

   Sample 2: _____

   Sample 3: _____

3. Explain your answer and the way that you figured it out.

Think!

There are 60 minutes in an hour. How many minutes are there in a day? How many days in a year?

Practice: Typing Tests

Over time, some people get better at skills while others do not. Below are four consecutive monthly typing test results for Kayla.

Month	Words Typed	Minutes	Minutes per Page (250 words)
January	100	5	
February	125	5	
March	200	8	
April	100	2	

1. Complete the last column in the chart above. Use the results for each test to predict how long it would have taken Kayla to type a page at this rate.

2. Tell the story of what happened to Kayla's typing skills over time. Include the number of words per minute during the time when she was the slowest. Also tell how many words per minute she typed at her fastest speed.

3. Use a diagram or the property of equal fractions to show whether Kayla's typing speed is the same for any two months.

Practice: Who Is the Fastest?

Several bikers have been chatting about their best race times. Each has won a race, although they have never competed against one another.

Each is claiming that he is the fastest biker. Look at their best times and then determine which of the racers is actually the fastest.

Biker	Race Length (in miles)	Best Time
Drake	50	2 hr. 30 min.
Leon	45	2 hr.
Denton	100	4 hr.

1. List the bikers from slowest to fastest. Give each of their speeds in miles per hour.

2. Show by diagram or the property of equal fractions why their rates are not the same.

Test Practice

1. It took Jai 30 minutes to roll 100 of his 1,000 newspapers. At this rate, how long would it take him to roll all of them?

   (1) 5 hours

   (2) 30 hours

   (3) 50 minutes

   (4) 100 minutes

   (5) 3,000 minutes

2. A stamping machine can stamp 500 letters in 3 minutes. At this rate, how long would it take the machine to stamp 10,000 letters?

   (1) 6 minutes

   (2) 50 minutes

   (3) 60 minutes

   (4) 1,500 minutes

   (5) 30,000 minutes

3. It took Marie 30 minutes to paint faces on 4 jack-o-lanterns. Which rate below is the same as Marie's?

   (1) 2 hours to paint 12 faces

   (2) 4 hours to paint 30 faces

   (3) 5 hours to paint 20 faces

   (4) 75 minutes to paint 10 faces

   (5) 400 minutes to paint 30 faces

4. A printing machine can print 250 business cards in 2 minutes. At this rate, how many cards can it print in 5 minutes?

   (1) 100

   (2) 500

   (3) 625

   (4) 1,250

   (5) 2,500

5. A bread slicer can slice 50 loaves of bread in 10 minutes. At this rate, how many loaves can it slice in 2 minutes?

   (1) 2

   (2) 5

   (3) 10

   (4) 25

   (5) 100

6. Johan lays bricks for patios. On average, he can lay 50 bricks in an hour. At this rate, how many hours would it take it to lay a patio that will need 275 bricks?

LESSON 3

Tasty Ratios

What do ratios have to do with taste?

You can adjust a recipe to make a larger or smaller amount of food than is called for by the recipe. When you change the amounts of ingredients in a recipe, it is important that you keep the ratios the same. If not, you may find that what you have created does not taste good or have the right texture.

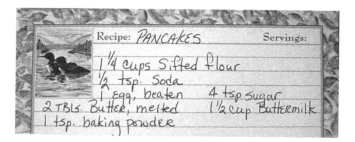

In this lesson, you will use your senses (taste and sight) to guess the ratios of ingredients in orange juice **mixtures**. You will also use diagrams and the property of equal fractions to show how the ingredients are (or are not) in proportion.

Activity 1: Orange Juice Taste Test

TO SERVE: Mix with 3 cans of cold water. Stir or shake briskly.

Sunrise Valley™
All Natural
100% PURE
ORANGE JUICE
FROZEN CONCENTRATE ℗
12 FL OZ (354mL) • PASTEURIZED • KEEP FROZEN

Here is the label from the can used to make the orange juice. Describe how each of the orange juice mixtures tastes and looks. What do you imagine the ratio of water to frozen **concentrate** is in each mixture?

Mixture A

1a. Describe the taste.

b. Describe the look.

c. What do you think the ratio of water to concentrate might be?

Mixture B

2a. Describe the taste.

b. Describe the look.

c. What do you think the ratio of water to concentrate might be?

Mixture C

3a. Describe the taste.

b. Describe the look.

c. What do think the ratio of water to concentrate might be?

Activity 2: "Doctoring" the Orange Juice

Look at the recipes that were used to make the three orange juice mixtures.

For each mixture, write the actual ratio of water to frozen concentrate.

Now that you know the actual ingredients, how would you change, or "doctor," Mixture B and Mixture C to taste like Mixture A without throwing any away?

Mixture	Actual Ratio of Water to Concentrate	How I Would "Doctor" the Mixture So It Tastes Like Mixture A
A		
B		
C		

Practice: How Does It Taste?

Look at the following proportions of ingredients in recipes. Place a check mark beside the description of how the adjusted recipe might turn out based on its new ratio.

1. *The Original Recipe*

   2 c. rice to 1 tsp. salt

   Too salty? _____   Just right? _____   Not salty enough? _____

   The Adjusted Version

   4 c. rice to 4 tsp. salt

2. *The Original Recipe*

   18 c. bread cubes to 3 eggs

   Too dry? _____   Just right? _____   Too wet? _____

   The Adjusted Version

   6 c. bread cubes to 1 egg

3. *The Original Recipe*

   2 T. lime juice to 3 T. sugar

   Too sweet? _____   Just right? _____   Too sour? _____

   The Adjusted Version

   6 T. lime juice to 6 T. sugar

4. *The Original Recipe*

   2 c. flour to $1\frac{1}{2}$ c. water

   Too dry? _____   Just right? _____   Too watery? _____

   The Adjusted Version

   5 c. flour to 3 c. water

5. *The Original Recipe*

   $\frac{3}{4}$ c. vinegar to 2 c. oil

   Too oily? _____   Just right? _____   Too vinegary? _____

   The Adjusted Version

   $1\frac{1}{2}$ c. vinegar to $3\frac{1}{2}$ c. oil

Practice: Lemonade for the Party

1a. How many cans of water are needed for each can of concentrate to make lemonade?

b. If you plan to serve each person 8 fl. oz. (1 cup), how many servings does one can make?

c. What is the ratio of water to frozen lemonade concentrate?

_____ water to _____ lemonade concentrate

2a. You have invited 15 people to the party. How many cans of frozen concentrate would you need to buy for the party?

b. How much water would you need to add in all?

c. How much lemonade would you have?

3. If you were adding ice cubes that would melt over time, what adjustments would you make to the recipe?

Practice: Building a Set of Equal Ratios

Create a set of ratios that are equal to the first one.

Example:

$$\frac{1}{3} = \frac{5}{15} = \frac{10}{30} = \frac{50}{150} = \frac{100}{300}$$

1. $\dfrac{2}{5} = \underline{\hspace{1.5em}} = \underline{\hspace{1.5em}} = \underline{\hspace{1.5em}} = \dfrac{60}{150}$

2. $\dfrac{1}{4} = \underline{\hspace{1.5em}} = \underline{\hspace{1.5em}} = \underline{\hspace{1.5em}} = \dfrac{200}{}$

3. $\dfrac{3}{4} = \underline{\hspace{1.5em}} = \underline{\hspace{1.5em}} = \underline{\hspace{1.5em}} = \dfrac{90}{}$

4. $\dfrac{4}{5} = \underline{\hspace{1.5em}} = \underline{\hspace{1.5em}} = \underline{\hspace{1.5em}} = \dfrac{120}{}$

Practice: Popcorn Party Mix

The recipe below makes about $2\frac{1}{2}$ quarts of party mix. How much of each ingredient would be needed for a triple batch?

1. Complete the following chart and then look at what happens to each of the numbers when you triple the amount.

Original Recipe	Triple the Recipe	Show How You Did It
2 qt. popcorn		
1 c. pretzels		
$1\frac{1}{2}$ c. cheese curls		
$\frac{3}{4}$ c. salted peanuts		
$\frac{1}{4}$ c. butter		
1 T. Worcestershire sauce		
$\frac{1}{2}$ tsp. garlic salt		

2. If the original recipe makes about $2\frac{1}{2}$ quarts of party mix, how many quarts would there be if you tripled the recipe?

Original Recipe	Triple the Recipe	Show How You Did It
$2\frac{1}{2}$ qt. party mix		

3. How does the amount of butter in the new recipe compare to the total amount of party mix in the new recipe? Show that the ratio is the same as in the original recipe.

Practice: Dressing It Up

Below are the basic ingredients for creamy garlic dressing.

1 c. mayonnaise

3 T. milk

2 T. cider vinegar

1 garlic clove

1a. Triple the recipe.

_____ c. mayonnaise

_____ T. milk

_____ T. cider vinegar

_____ garlic cloves

b. What is the ratio of mayonnaise to milk in the original recipe?

c. What is the ratio of mayonnaise to milk in the new recipe?

2a. Quadruple the original recipe.

_____ c. mayonnaise

_____ T. milk

_____ T. cider vinegar

_____ garlic cloves

b. What is the ratio of milk to vinegar in the original recipe?

c. What is the ratio of milk to vinegar in the quadrupled recipe?

Practice: Doctor This

Following are three stories about individuals who have problems with their recipes. Think about how you can help them to readjust their ingredients so that everything is still in balance.

1. A recipe for a cake calls for half a stick of butter and two eggs. Ingram was distracted and beat in 1 stick of butter and three eggs.

   a. What should he do now to fix the proportions of the ingredients?

   b. How will he have to adjust the remaining ingredients?

   c. How many cakes will he end up making?

2. A brownie recipe calls for 2 cups of sugar and 3 cups of flour. Rachel discovers that she only has 1 cup of sugar.

   a. How could she adjust the recipe to keep the ingredients in balance?

   b. How will she have to adjust the remaining ingredients?

3. A recipe that makes 5 dozen cookies calls for 3 eggs and 6 cups of flour. Sue only has 2 eggs.

   a. How could she adjust the recipe to keep the ingredients in proportion?

   b. What will she have to do with the remaining ingredients?

   c. About how many cookies will she end up making?

Extension: Reasoning with Ratios

You can make punch with strawberry syrup and water.

1. Pitcher A contains punch that has a stronger strawberry taste than the punch in Pitcher B. If one cup of strawberry syrup is added to Pitcher A and one cup of water is added to pitcher B, which pitcher will contain the punch with the stronger strawberry taste? Explain your reasoning.

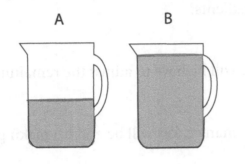

A   B

2. The punch in both Pitchers A and B has the same taste. If one cup of strawberry syrup is added to both pitchers, which pitcher will contain the punch with the stronger strawberry taste? Explain your reasoning.

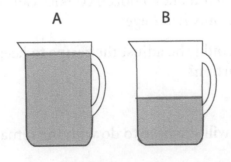

A   B

3. Pitcher A contains punch with a stronger strawberry taste. If one cup of strawberry syrup is added to both pitchers, which pitcher will contain the punch with the stronger strawberry taste? Explain your reasoning.

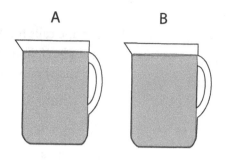

A   B

4. Pitcher B contains punch with the stronger strawberry taste. If one cup of syrup is added to Pitcher A and one cup of water is added to Pitcher B, which pitcher will contain the punch with the stronger strawberry taste? Explain your reasoning.

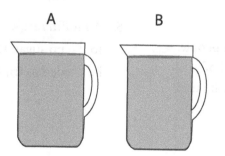

A   B

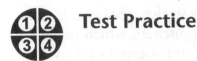

Test Practice

1. A recipe calls for 2 cups of sugar to $\frac{1}{2}$ cup of lime juice. How much lime juice should be added if the new recipe uses 6 cups of sugar?

   (1) 1 cup

   (2) $1\frac{1}{2}$ cups

   (3) 2 cups

   (4) $2\frac{1}{2}$ cups

   (5) 3 cups

2. A recipe calls for $\frac{1}{2}$ cup of vinegar to $\frac{3}{4}$ cup of oil. How much vinegar should be added if the new recipe uses $1\frac{1}{2}$ cups of oil?

   (1) $\frac{1}{4}$ cup

   (2) $\frac{1}{2}$ cup

   (3) $\frac{3}{4}$ cup

   (4) 1 cup

   (5) $1\frac{1}{2}$ cups

3. A recipe calls for 3 potatoes to serve 4 people. If Jacques is planning to serve a crowd of 20 people, how many potatoes should he use?

   (1) 7 potatoes

   (2) 12 potatoes

   (3) 15 potatoes

   (4) 60 potatoes

   (5) 80 potatoes

4. A recipe calls for 3 cups of water to serve 8 people. If Shaina is planning to serve 4 people, how many cups of water should she use?

   (1) 1 cups

   (2) $1\frac{1}{2}$ cups

   (3) 2 cups

   (4) 7 cups

   (5) 12 cups

5. A party punch calls for 6 oz. of concentrate to 5 cups of water. If you only have 3 oz. of concentrate, how much water should you use?

   (1) 2 cups

   (2) $2\frac{1}{2}$ cups

   (3) 3 cups

   (4) $3\frac{1}{2}$ cups

   (5) 4 cups

6. A muffin recipe calls for 3 tsp. of baking powder to $\frac{1}{3}$ cup sugar. How many cups of sugar would be needed to triple the recipe?

Another Way to Say It

How are ratios used to influence others?

News reports and advertisements use **percents**, fractions, and ratios to communicate and persuade. Because advertising claims can be tricky to understand, it is important to think carefully about what the statements mean.

This lesson will give you a chance to build on what you know about making comparisons between two amounts. First you will work with the amounts of concentrate, water, and total mixture from the lesson on taste-testing orange juice. Then you will think about other ways to write comparisons, including those for advertisements.

Activity 1: Part to Part versus Part to Whole

1. Complete the following table by writing and drawing comparisons of two of the three amounts of ingredients (water, concentrate, and orange juice mixture) at a time. Use the recipe with the correct ratios.

Two Ingredients Being Compared	Ratio of the Amounts of Two Ingredients	Picture of the Amounts of Two Ingredients
Example: Water to mixture	3:4	⬜⬜⬜⬛

2. Which of the comparisons above uses **part to part** ratios?

3. Which of the comparisons above uses **part to whole** ratios?

Activity 2: Two Truths and a Lie

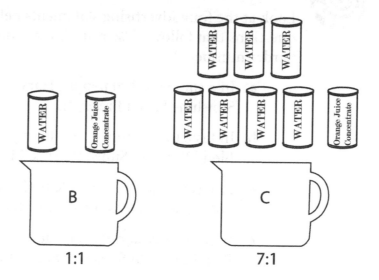

Think about the comparisons of the amounts of ingredients used to create Mixture B and Mixture C.

1. Write two truths and a lie for Mixture B and/or Mixture C. At least one of your statements should use a percent.

2. When you are done, pair up with another student and exchange your statements. Try to figure out which are the truths and which is the lie. Use pictures to illustrate why some statements are true and some are false.

Activity 3: Making Advertising Statements

Look at the four advertising statements below and answer the questions that follow. (Use rounding to make it easier to think about harder numbers.)

a. In a survey of 100 people, Harvest Blend Bread was preferred to Sunshine Sliced Bread, 4:1.

b. In taste tests, about 80% of the people polled preferred Harvest Blend Bread to Sunshine Sliced Bread.

c. In taste tests, 5,975 people preferred Harvest Blend Bread and only 1,486 preferred Sunshine Sliced Bread.

d. In taste tests, about four out of five people preferred Harvest Blend Bread to Sunshine Sliced Bread.

1. Use a diagram or picture to describe to someone else what each statement means.

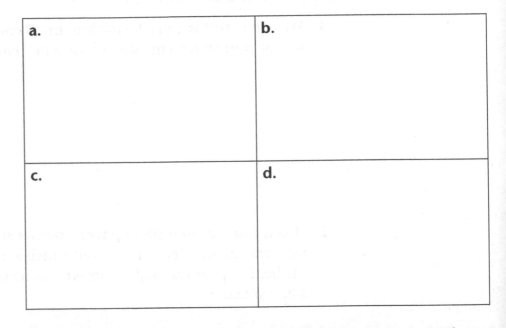

2. Look at your diagrams. Do all four send the same message? Why or why not?

3. Which of the four statements would you use to sell Harvest Blend Bread? Why?

Practice: What Is Another Way to Say It?

In a survey of 100 people, 7 out of 10 people preferred Misty Mouthwash to Refresh Mouth Rinse.

1. Write another ad that says the same thing another way.

2. Explain how you know that the two statements say the same thing.

3. Share your ad and the original with a friend. Does your friend think the two ads say the same thing?

Practice: Keeping the Ratio the Same with Gardeners

Gardeners preferred Snail Away to Slug-It-Out by a ratio of 9 to 6.

1. Give three other sets of ratios of gardeners' preference for Snail Away that still show a ratio of 9 to 6.

2. What if a total of 450 gardeners were asked to compare Snail Away and Slug-It-Out? About how many of them would prefer Snail Away?

Practice: Who Is Right?

Students in an adult education class conducted a survey to determine whether they wanted to hold classes on Saturday. Fifteen students said they would like to hold class on Saturday; 25 said they would *not* like Saturday classes.

Caroline then reported that only 3 out of 8 students wanted Saturday classes.

Bronson argued that the ratio of those wanting Saturday classes to those who did not was 3 to 5.

Who is right? Support your answer with drawings or descriptions.

Practice: Rounding Ratios

Round each of the set of comparisons below to make them more workable.
Example:

Think: 6,915 rounded is 7,000;
10,345 rounded is 10,000.

6,915 doctors out of 10,345   7,000 out of 10,000 doctors

1. 8,132 dentists compared to 3,921 dentists

2. 7,131 teachers out of 8,099 teachers

3. 3,871 campers compared to 9,821 campers

4. 6,889 renters out of 9,175 renters

5. 924 managers compared to 4,214 managers

6. 9,231 truck drivers out of 15,211 truck drivers

7. 4,781 teens out of 18,970 teens

 **Test Practice**

1. According to a survey, people preferred Snick Snacks to potato chips by a ratio of 7 to 3. Which of the following is another way to make the same comparison?

   (1) 7 out of 10 people prefer Snick Snacks to potato chips.

   (2) 4 out of 7 people prefer Snick Snacks to potato chips.

   (3) 3 out of 7 people prefer Snick Snacks to potato chips.

   (4) 30% of the people prefer Snick Snacks to potato chips.

   (5) 3 out of 10 people prefer Snick Snacks to potato chips.

2. According to a taste test, people who preferred iced tea outnumbered those preferring hot tea by a ratio of 1,232 to 597. Which of the following is another way to make the same comparison?

   (1) About 50% of the people preferred iced tea to hot tea.

   (2) About 6 out of 12 people preferred ice tea to hot tea.

   (3) About 2 out of 3 people preferred ice tea to hot tea.

   (4) About 6 out of 18 people preferred ice tea to hot tea.

   (5) About 1 out of 2 people preferred ice tea to hot tea.

3. According to a survey, 6 out of 10 adults prefer reading mysteries to comedies. Which of the following is another way to make the same comparison?

   (1) Adults prefer mysteries to comedies by a ratio of 5 to 3.

   (2) Adult prefer mysteries to comedies by a ratio of 2 to 3.

   (3) Adults prefer mysteries to comedies by a ratio of 6 to 10.

   (4) 40% of those surveyed prefer mysteries to comedies.

   (5) 60% of those surveyed prefer mysteries to comedies.

4. According to a survey, people preferred reading the newspaper in the evening rather than the morning by a ratio of 8 to 2. Which of the following is *not* another way to make the same comparison?

   (1) 80% of the people surveyed would rather read the newspaper in the evening.

   (2) Two out of 10 people surveyed would rather read the newspaper in the morning.

   (3) Eight out of 10 people surveyed would rather read the newspaper in the evening.

   (4) One out of 5 people surveyed would rather read the newspaper in the evening.

   (5) One out of 5 people surveyed would rather read the newspaper in the morning.

5. According to a taste test, the ratio of people who preferred ice cream to those who preferred frozen yogurt was 803 to 799. Which of the following is *not* another way to make the same comparison?

(1) About 50% of the people preferred ice cream to frozen yogurt.

(2) About 1 out of every 2 people preferred ice cream to frozen yogurt.

(3) About 8 out of 16 people preferred ice cream to frozen yogurt.

(4) About 1 out of every 8 people preferred ice cream to frozen yogurt.

(5) About 1 out of every 2 people preferred frozen yogurt to ice cream.

6. Out of 50 people surveyed, only 20 said that they did not like the new shopping hours at the mall. What percent of those surveyed were *not* opposed to the new shopping hours?

Mona Lisa, Is That You?

> **Do the proportions appear correct?**

Just as the ingredients in a recipe need to be kept in proportion, so do the dimensions of a picture. When the proportions get distorted, things just don't look right.

Do you have a good sense of visual balance? In this lesson, you will "eyeball" a set of art reproductions to judge which are good and which are distorted. Then you will test your judgment with actual measurements. Finally, you will use a graph as a tool for examining whether ratios are equal.

Activity 1: How Large Is the *Mona Lisa* Painting?

The *Mona Lisa*, painted by Leonardo da Vinci from 1503 to 1506, is one of the most famous paintings in the Western world. Some people think it is her mysterious smile that makes her so intriguing.

Below is a good miniature reproduction of *Mona Lisa*. In actual size, the length, or height, of the original painting is about 30 inches.

1. What is the actual width?

2. How do you know?

3. Draw the actual size of the *Mona Lisa* on a large piece of paper.

Activity 2: Mona Lisa, Is That You?

Part 1

A graphic artist was experimenting on his computer trying to make some reproductions of da Vinci's *Mona Lisa*. Sometimes he succeeded and other times the results were distorted. His first attempt (A) turned out fine.

1. Use your own sense of proportion to sort the seven pictures of *Mona Lisa* into two piles, good ones and distorted ones. Record each below.

Good Reproductions	Bad Reproductions

2. Now use measurements and numbers to test the accuracy of your judgment about which are good and which are distorted reproductions. Measure the width and height of each of the choices and record them below.

Good Reproductions	Bad Reproductions

Part 2

1. Next create a set of equal ratios with the dimensions of the good reproductions.

$$\frac{\text{width (in inches)}}{\text{height (in inches)}} \quad \frac{2}{3} = \underline{\hspace{2cm}} = \underline{\hspace{2cm}} = \underline{\hspace{2cm}}$$

2. What patterns do you see in these numbers?

3. Now look at your list of distorted reproductions. Show that each of the distorted reproduction's dimensions is *not* in a 2 to 3 ratio.

4. Change each of those distorted reproductions into good ones by changing only one of the dimensions—length or width. Include an explanation of how you made the change.

Activity 3: Lining Up the Reproductions

Part 1

The Good Reproductions

Line up the lower left-hand corners of all the good reproductions, including the original and the 20″ by 30″ reproduction you drew on the newsprint.

1. Trace the outline of each picture on the newsprint with the 20″ by 30″ reproduction, always beginning at the lower left-hand corner.

2. Mark the upper right-hand corner of each picture with a ✳.

3. What do you notice?

4. Why do you think that is?

The Distorted Reproductions

Now line up the distorted reproductions with the lower left hand corners together. With another color pen or marker, trace the bad reproductions on the same large piece of paper.

1. Trace the outline of each picture.

2. Mark the upper right-hand corner of each with an "X."

3. What do you notice?

4. Why do you think that is?

Part 2

Now graph the height versus the width of your pictures on the grid below.

1. Mark the good reproductions with a ✶ and the distorted ones with an "X" in another color.

2. What patterns do you notice?

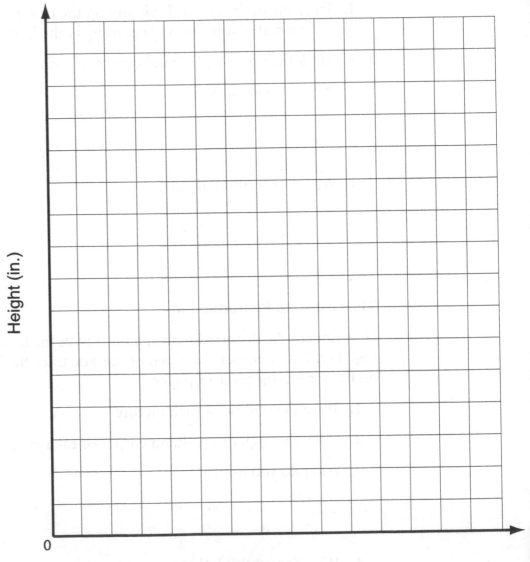

Height (in.)

0

Width (in.)

Practice: Rectangle Ratios

Put a check beside each of the shapes that has the same proportion as the shape below.

1. _____ a rectangle that is 2 inches high and 3 inches wide

2. _____ a rectangle that is 3 inches high and 6 inches wide

3. _____ a rectangle that is 3 inches high and 1 inch wide

4. _____ a rectangle that is 2 inches high and 6 inches wide

5. _____ a rectangle that is 4 inches high and 12 inches wide

6. _____ a rectangle that is 6 inches high and 3 inches wide

7. _____ a rectangle that is 3 inches high and 9 inches wide

8. Summarize by making a list of equal ratios.

Practice: What Is the Pattern?

1. What is the pattern used to generate each set of equal ratios below?

   Example:

   $$\frac{3 \text{ inches}}{4 \text{ inches}} = \frac{6 \text{ inches}}{8 \text{ inches}} = \frac{12 \text{ inches}}{24 \text{ inches}}$$ *double, or multiply by 2*

   a. $\dfrac{7 \text{ feet}}{10 \text{ feet}} = \dfrac{21 \text{ feet}}{30 \text{ feet}} = \dfrac{63 \text{ feet}}{90 \text{ feet}}$

   b. $\dfrac{40 \text{ cm}}{50 \text{ cm}} = \dfrac{20 \text{ cm}}{25 \text{ cm}} = \dfrac{10 \text{ cm}}{12.5 \text{ cm}}$

   c. $\dfrac{2 \text{ feet}}{3 \text{ feet}} = \dfrac{8 \text{ feet}}{12 \text{ feet}} = \dfrac{32 \text{ feet}}{48 \text{ feet}}$

   d. $\dfrac{300 \text{ cm}}{500 \text{ cm}} = \dfrac{30 \text{ cm}}{50 \text{ cm}} = \dfrac{3 \text{ cm}}{5 \text{ cm}}$

   e. $\dfrac{36 \text{ inches}}{144 \text{ inches}} = \dfrac{6 \text{ inches}}{24 \text{ inches}} = \dfrac{1 \text{ inch}}{4 \text{ inches}}$

2. Make up your own set of equal ratios. Then ask a classmate to figure out what your pattern is.

Practice: Using a Graph to Predict

1. Imagine that you make $0.10 profit for every $1.00 of merchandise sold. Create a graph showing your rate of profit. Remember that $0 sold equals $0 profit.

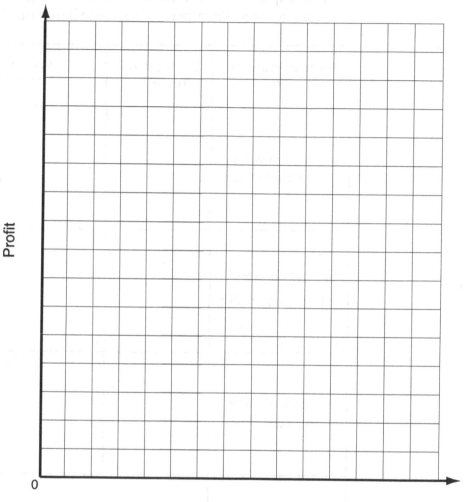

Profit

0

Selling Price

2. Based on your graph, predict what your profit would be if you sold $100 worth of merchandise.

3. Put a check beside each of the following statements that you predict would have the same profit rate as yours:

   a. _____ $7 sold, profit of $0.07

   b. _____ $7 sold, profit of $0.35

   c. _____ $7 sold, profit of $0.70

4. Check your answers for Problem 3 by adding those profit rates to your graph above.

Practice: Graphing Ingredients

A recipe for rice calls for two cups of water for every one cup of rice. Use this ratio to guide your reasoning below.

1. Complete the table below by determining which of the ratios below are correct. Then predict the taste based on the ratio.

Ratio of Water to Rice	Correct Ratio? Yes or No	What Do You Think the Recipe Will Taste Like?
a. 3 cups of water to 2 cups of rice		
b. 5 cups of water to $2\frac{1}{2}$ cups of rice		
c. 6 cups of water to 3 cups of rice		
d. 7 cups of water to 3 cups of rice		
e. 9 cups of water to $4\frac{1}{2}$ cups of rice		

2. Now graph all the points and see whether your predictions were correct.

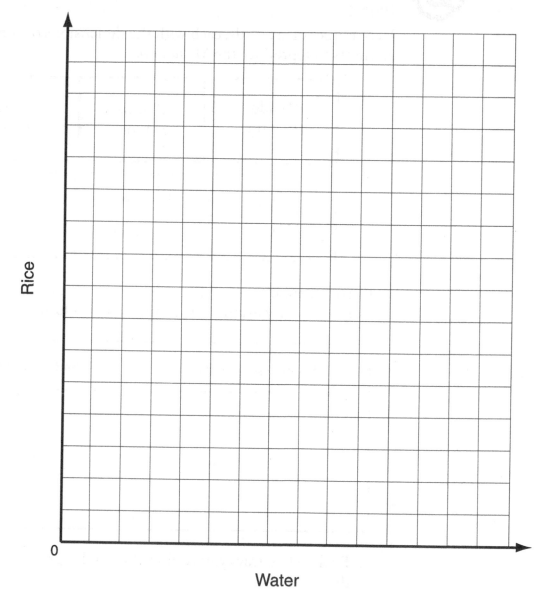

3. How does your graph compare to your predictions?

Extension: Paint By Numbers

Part 1

Here is a set of six cards, each with the dimensions of some more attempts to reproduce the *Mona Lisa*.

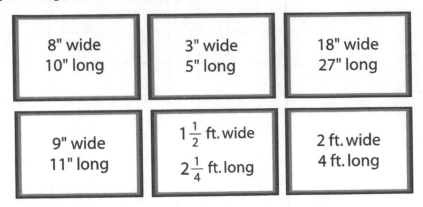

8" wide 10" long	3" wide 5" long	18" wide 27" long
9" wide 11" long	$1\frac{1}{2}$ ft. wide $2\frac{1}{4}$ ft. long	2 ft. wide 4 ft. long

1. Sort them into two piles: good reproductions and distorted ones. List the dimensions of the cards in the columns below.

Good Reproductions	Bad Reproductions

2. Explain two ways you could tell a good reproduction from a distorted one.

Part 2

1. Now graph the height versus the width of the six pictures on the grid below. Mark the good reproductions with a ✳ and distorted ones with an "X" in another color.

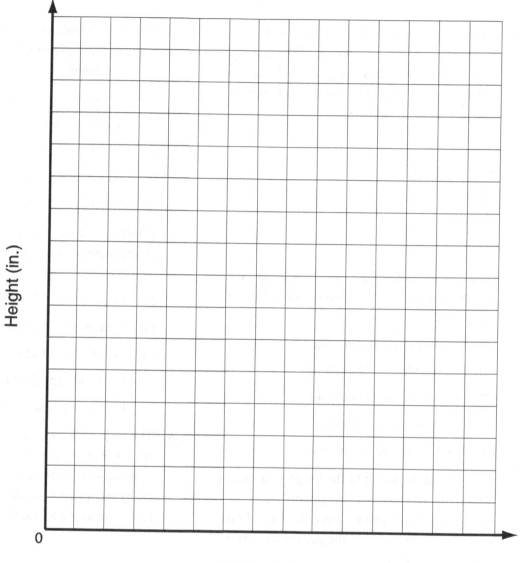

Width (in.)

2. What patterns do you notice?

1. The ratio of the width to the length of a rectangle is 3 inches to 5 inches. If the rectangle were enlarged so that its length were 30 inches, how many inches wide would the rectangle need to be to still be in balance?

    (1) 8

    (2) 15

    (3) 18

    (4) 50

    (5) 90

2. Which of the following would be a distorted enlargement of a picture that is 4 cm wide and 8 cm long?

    (1) 8 cm wide and 16 cm long

    (2) 10 cm wide and 20 cm long

    (3) 12 cm wide and 24 cm long

    (4) 16 cm wide and 8 cm long

    (5) 20 cm wide and 40 cm long

3. The ratio of the width to the length of a picture is 4 inches to 6 inches. If the picture were enlarged so that its width were 20 inches, how many inches long would the picture have to be to still be in balance?

    (1) 10

    (2) 24

    (3) 30

    (4) 80

    (5) 120

4. Which of the following would be a distorted reduction of a picture that is 36 inches wide and 48 inches long?

    (1) 2 inches wide and 12 inches long

    (2) 3 inches wide and 4 inches long

    (3) 6 inches wide and 8 inches long

    (4) 9 inches wide and 12 inches long

    (5) 18 inches wide and 24 inches long

5. Which of the following would be a distorted enlargement of a picture that is 2 cm wide and 5 cm long?

    (1) 4 cm wide and 10 cm long

    (2) 5 cm wide and 12.5 cm long

    (3) 6 cm wide and 15 cm long

    (4) 7 cm wide and 20 cm long

    (5) 10 cm wide and 25 cm long

6. The ratio of the width to the length of a picture is 12 inches to 18 inches. If the picture were reduced so that its width were 4 inches, how many inches long would the picture have to be to still be in proportion?

Redesigning Your Calculator

How do proportions affect design choices?

Ratios play an important role in design. Architects, graphic designers, engineers, and artists call upon their sense of proportion in their work. In this lesson, you will measure and draw a life-size version of your calculator, then larger and smaller versions.

You will explore the relationship between the **dimensions** and the size of the surface, or **area**, as you figure out the answers to these two questions:

- If you double the dimensions, will you need twice the amount of material?

- If you halve the dimensions, will you need half the amount of material?

Activity 1: Three Calculators

Have your calculator by your side for this activity. Not only will you use it to check your math, but you will also redesign the calculator itself!

The Large-Print Version

The company that manufactures your calculator wants to produce another similar model for people whose eyesight requires larger print.

The company asks you to design a calculator that is double the dimensions of your original model in length and width. They want you to keep the other features—the display window and the buttons—in proportion.

The Wallet-Size Version

The company that manufactures your calculator also wants to provide a model for people who want to be able to tuck the calculator into their wallets.

They ask you to design a calculator that has dimensions half of your original's in length and width. Again, they want you to keep the display window and buttons in proportion.

On grid paper, make the following drawings and label their dimensions:

1. Your calculator (actual size without the cover, if it has one)

2. The large-print version of your calculator

3. The wallet-size version of your calculator

Include in all the drawings: the outline of the calculator, the display window, and at least one of the numbered buttons. Then record all measurements in the table on page 73. Measure in centimeters (cm) to the nearest 0.1 cm.

The Measurements

1. Record all the measurements of the three calculators in decimal form to the nearest 0.1 cm.

What to Measure (in cm)	Your Calculator	The Large-Print Version	The Wallet-Size Version
a. Calculator width			
b. Calculator height			
c. Display window width			
d. Display window height			
e. Numbered button width			
f. Numbered button height			

2. Look across the rows. What patterns do you see?

3. List some examples of equal ratios that you see in your table above.

Activity 2: Material for a Calculator Cover

The company that manufactures these new sizes of calculators wants to provide protection for them. It would like to develop a clear plastic cover for the top of the calculator.

First estimate the amounts of plastic needed for the different models of the calculator and then determine the exact amounts. Use the following chart for your answers.

1. Estimate the amount of plastic you would need for the top cover of each of the sizes of calculators.

2. Determine the amount of plastic you would need for the top cover of each of the sizes of calculators. How do these actual amounts compare with your predictions?

What to Measure	Your Calculator	The Large-Print Version	The Wallet-Size Version
a. Calculator width (in cm)			
b. Calculator height (in cm)			
c. Estimate of area (in sq. cm)			
d. The actual area (in sq. cm)			
e. The perimeter (in cm)			

Look for patterns in the table to answer the following questions:

3. When you double the dimensions, does the amount of material needed for the cover double?

4. Explain your reasoning.

5. When you halve the dimensions, do you need half the amount of plastic?

6. Explain your reasoning.

7. Compare the amount of plastic needed for the large-print calculator with the amount needed for the wallet-size calculator.

8. When you double the dimensions, how does the **perimeter** (the distance around the edge) change?

9. When you double the dimensions, how does the area change?

Practice: Graphing Calculator Information

Line up the lower left-hand corners of each of the three calculator drawings from *Activity 1*.

Make a large piece of grid paper by taping two pieces together. Trace the outline of each calculator on grid paper.

Mark the upper right hand corner of each of the three calculators with a ✳.

1. What do you notice?

2. Why do you think that is?

Practice: Similar Rectangles

Shapes that are **similar** have side measurements that are in the same proportion to one another. You can often eyeball shapes to tell whether they are in proportion or not.

1. First classify the shapes by sight. Which do you think are similar?

2. Now measure and label the sides. Are the longer and shorter sides of each group of similar shapes in the same proportion? How do you know?

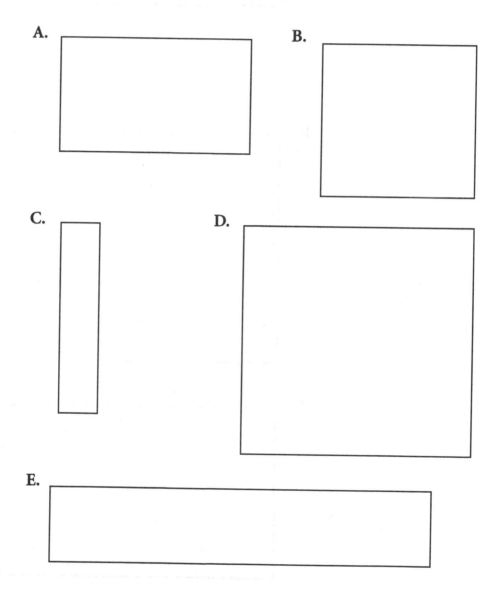

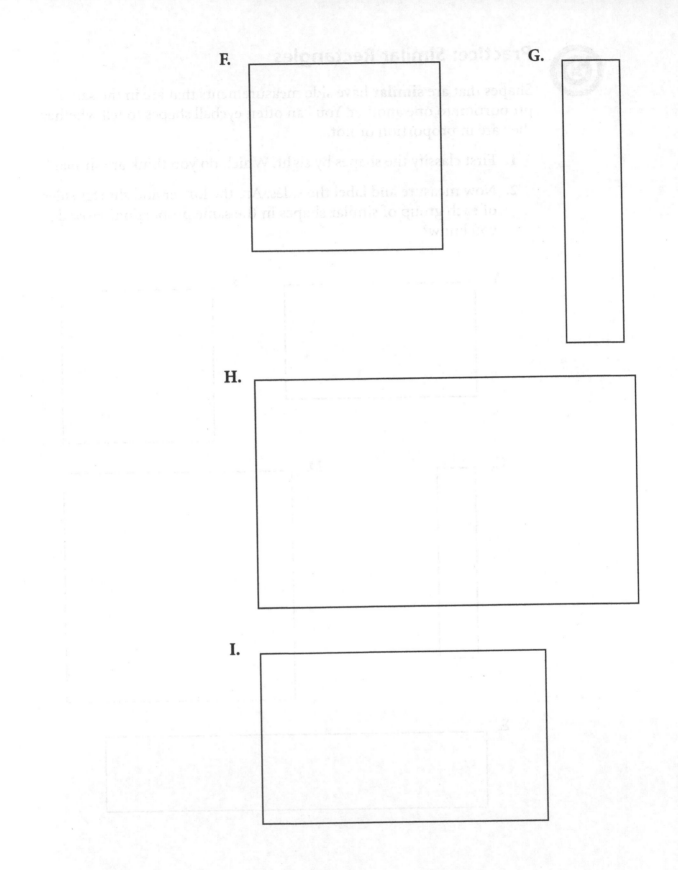

F.

G.

H.

I.

3. On the following grid, graph the longer side versus the shorter side of each rectangle. Do the points land where you would expect them? Explain.

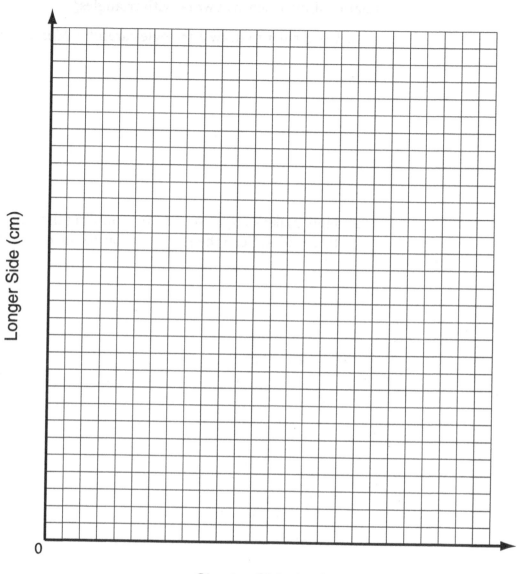

Longer Side (cm)

Shorter Side (cm)

0

Practice: Shaping Up!

You have been working with similar rectangles. What happens to the relationships when you work with triangles?

1. Use a ruler to draw a triangle carefully. Measure and label the sides.

2. Now make a larger version by doubling the length of each of the three sides. Keep the shape similar to the original.

3. About how many of the smaller triangles appear to cover the surface of the larger triangle?

4. What happened to the size of the angles?

Practice: Squaring Off

A day-care provider planned to have a corner of her basement tiled. At first she planned to have a square with a length of 5 feet. Then she decided to enlarge the square so that it would be 10 feet in length.

1. What is the ratio of the original length to the new length?

2. Is the ratio of the original area to the new area the same ratio as the lengths? Explain your answer using pictures or descriptions.

3. How many square tiles for the new area would she need if they each measure 12″ by 12″?

4. How many times more tiles is this than she would have needed for her original square? Use a drawing to support your answer.

Extension: Perimeter, Surface Area, and Volume

You have to pack several objects into a rectangular shaped box. You suggest to your coworker that if you double the dimensions of the box, you could double the amount of room for packing.

You coworker scratches his head and wonders whether it would also require twice as much material to make the box.

The two of you decide to test the idea. Using the box below as your original, create a box that is twice as long, twice as wide, and twice as tall. Then see what happens to the measurements of the box.

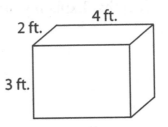

1. Record the measurements in the following table. Predict the **surface area** and actual measurements, then do the same for volume.

What to Measure	Original Box Measurements	Estimation for New Size Box	Actual Measurements
a. Length			
b. Width			
c. Height			
d. Surface Area			
e. Volume			

2. How are the length, width, and height of the new box different from the original?

3. What happened to the surface area of the original box when you doubled the dimensions? Why? How will this impact the usefulness of the new shape?

4. What happened to the volume of the box? Why? How will this impact the usefulness of the new shape?

Questions 1 and 2 refer to the following graph:

Calculator Preferences

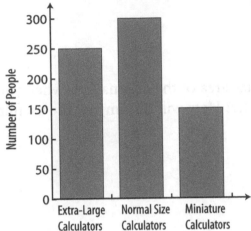

1. According to the bar graph, what is the ratio of the number of people who prefer miniature calculators to those who prefer extra-large calculators?

   (1) 1:2

   (2) 2:1

   (3) 3:5

   (4) 30:25

   (5) 15:70

2. According to the bar graph, what is the ratio of the number of people who prefer normal size calculators to those who prefer extra-large calculators?

   (1) 25:70

   (2) 15:25

   (3) 25:15

   (4) 30:25

   (5) 30:70

3. Jacobia drew a square having sides of 4 cm each. Next she drew a shape that was double the length and width of the square. What does her new shape look like?

   (1) Square with a perimeter of 8 cm

   (2) Square with a perimeter of 32 cm

   (3) Rectangle with a perimeter of 8 cm

   (4) Rectangle with a perimeter of 24 cm

   (5) Rectangle with a perimeter of 16 cm

4. Each of the sides of a logo that was 2 inches by 4 inches was tripled in size for a poster. How many inches of trim would be needed to surround the new size?

   (1) 18

   (2) 24

   (3) 28

   (4) 36

   (5) 48

5. Each of the sides of a logo that was 2 inches by 4 inches was tripled in size for a poster. What is the area of the new poster?

   (1) 8 sq. in.

   (2) 18 sq. in.

   (3) 24 sq. in.

   (4) 36 sq. in.

   (5) 72 sq. in.

6. A company manufactured a transistor radio that was one-fourth the size of its regular transistor radio but similar in shape. If the regular radio had a length of 12 inches and a width of 8 inches, how many inches long is the new transistor radio?

Comparing Walks

> *What's our average walking speed?*

In our fast paced modern life, speed is often a concern, whether we're flying in a plane, driving a car, jogging, or walking. Do you know about how fast you walk? How does this compare to the rate at which your classmates walk?

In this lesson, you will find your walking rate and then compare your rate to those of your classmates. You will also practice comparing ratios with a useful tool: the **unit rate**.

Activity 1: Team Walking Speeds

List the names of your team members:

How fast does each person on your team walk at his or her normal speed? Who walks the fastest? With your team, agree upon a way to figure out each person's normal walking speed.

Prepare a presentation so everyone can follow your team's reasoning.

Include the following:

1. A table with the information your team collects.

2. A list with teammate's walking speeds arranged from fastest to slowest.

3. An explanation of how you determined who was the fastest.

4. The average walking speed for your team and an explanation of how you figured it out.

Challenge:

5. At these walking rates, how long would it take each person on your team to walk a mile?

Activity 2: Class Walking Speeds

Once you and the rest of your class decide on a common way to represent the rate of walking, work with your team to convert all the rates to the same format. Copy everyone's rate from the presentations. Use a calculator to find the unit rate for each. Be sure to clearly label the distance and time units so that you know what each represents.

Name	Walking Speed	Unit Rate

1. Who walked the fastest? How do you know?

2. Who walked the slowest? How do you know?

3. What is the average walking speed for the class? Explain your reasoning.

Practice: Rounding Decimals from the Calculator Screen

A calculator is very helpful when doing arithmetic. However, the answer may have more decimal places than you need for your purposes.

Rounding to the nearest hundredth (0.01)

Try dividing $2.00 by 3 on the calculator. On some calculators, the answer appears as 0.666666666. When working with dollars and cents, we usually round to the nearest cent, or to the hundredths place. So, an answer rounded to the nearest cent (or hundredth) would be more workable as $0.67.

Use a calculator to perform the division. First, write the answer exactly as it appears on the screen. Then make the number more manageable by rounding it to the nearest hundredth.

	Answer on the calculator screen	Answer rounded to the nearest hundredth
1. 1 ÷ 3	a. _____	b. _____
2. 5 ÷ 8	a. _____	b. _____
3. 17 ÷ 21	a. _____	b. _____
4. 150 ÷ 62	a. _____	b. _____
5. 62 ÷ 1,000	a. _____	b. _____
6. 1,000 ÷ 62	a. _____	b. _____

Rounding to the nearest tenth (0.1)

Sometimes, rounding to one decimal place, or to the nearest tenth is close enough for a workable number. Round each of the six numbers above to the nearest tenth.

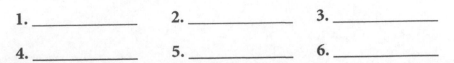

1. _____   2. _____   3. _____

4. _____   5. _____   6. _____

Rounding to the nearest whole number

Sometimes, rounding to the nearest whole number is close enough. Round each of the six numbers above to the nearest whole number.

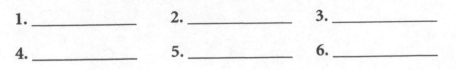

1. _____   2. _____   3. _____

4. _____   5. _____   6. _____

Practice: Converting to the Unit Rate

Change each of the following rates to the unit rate. Use labels to describe what that unit rate represents. Do the division in your head when you can, and use a calculator when the numbers get more difficult.

Example:

Given Rate

$$\frac{20 \text{ yd.}}{30 \text{ sec.}}$$

Unit Rate

0.667 yd./sec.

Given Rate   **Unit Rate**

1. $\dfrac{18 \text{ yd.}}{18 \text{ sec.}}$

2. $\dfrac{30 \text{ ft.}}{12 \text{ sec.}}$

3. $\dfrac{10 \text{ yd.}}{3 \text{ sec.}}$

4. $\dfrac{50 \text{ yd.}}{3 \text{ min.}}$

5. $\dfrac{12 \text{ in.}}{10 \text{ sec.}}$

6. $\dfrac{14 \text{ ft.}}{20 \text{ sec.}}$

7. $\dfrac{45 \text{ ft.}}{60 \text{ sec.}}$

8. Which of the above rates is the fastest?

Practice: Who Is the Fastest?

For each set of races, determine who was the fastest. You might want to make a table and change each rate to a unit rate so you can compare them easily.

1. Sprint

Joe clocked himself at 24 feet in 8 seconds, Seth timed himself at 25 feet in 10 seconds, and Jorge clocked himself at 30 feet in 14 seconds.

2. Walk for Humanity

These women described their times in the Walk for Humanity:

     Shondra: 5 miles in 2 hours

     Tabitha: 3 miles in 60 minutes

     Narveen: 3.5 miles in 1 hour 30 minutes

     Ashlyn: 5 miles in 150 minutes

3. Swim Meet

The following times were recorded:

     Chee: 5 minutes for 6 laps

     Marsha: 8 minutes for 10 laps

     Morgan: 6 minutes for 8 laps

Practice: MPH, MPG, and Other Rates

Use your estimation skills to approximate each of the following rates.

Example: 48 mph is about _200_ miles in 4 hours.

1. 148 miles in two hours is about _____ miles in 6 hours.

2. 105 calories in one hour is about _____ calories in $2\frac{1}{2}$ hours.

3. 36 mpg (miles per gallon) is about _____ gallons for 350 miles.

4. 8 hours to travel 650 miles is about _____ mph.

5. 4 days to bike 310 miles is about _____ miles per day.

6. 74 mph is about _____ miles in $3\frac{1}{2}$ hours.

7. 992 miles in 3 days is about _____ miles per day.

8. 348 calories in 2 hours is about _____ calories in 3 hours.

Extension: On the Treadmill

A woman set the speed on a gym treadmill at 3 miles per hour, kept the incline flat, and then pressed "Start." The treadmill started moving, and she began to walk at that constant speed.

Below is a "snapshot" of the digital display on the treadmill a few moments after she began to walk.

Snapshot 1

Calories	Distance (mi.)	Time (min:sec)	Incline	Speed (mph)
10	0.13	2:45	0.0	3.0

A while later, a second snapshot showed the following information:

Snapshot 2

Calories	Distance (mi.)	Time (min:sec)	Incline	Speed (mph)
20	0.25	5:15	0.0	3.0

1. What is the same?

2. What has changed?

3. Do the changes make sense? Why?

4. In Snapshot 3, what would you approximate the display to look like at the 10-minute mark?

Snapshot 3

Calories	Distance (mi.)	Time (min:sec)	Incline	Speed (mph)
		10:00	0.0	3.0

5. In Snapshot 4, fill in what you would expect the display to read (approximately) when the woman had walked a mile?

Snapshot 4

Calories	Distance (mi.)	Time (min:sec)	Incline	Speed (mph)
	1.0		0.0	3.0

6. At this rate, about how long would it take to walk off 250 calories?

Snapshot 5

Calories	Distance (mi.)	Time (min:sec)	Incline	Speed (mph)
250			0.0	3.0

7. In Snapshot 6, show how long it would take to walk 4.5 miles at this rate. How do you know?

Snapshot 6

Calories	Distance (mi.)	Time (min:sec)	Incline	Speed (mph)
	4.5		0.0	3.0

Extension: Graphing Treadmill Distances versus Times

On the grid below, plot the distances and times from the six treadmill readings in *Extension: On the Treadmill*, pages 92–93. Use the graph to predict how far the woman on the gym treadmill would walk in 45 minutes. Plan carefully so that you fit all the points on the graph.

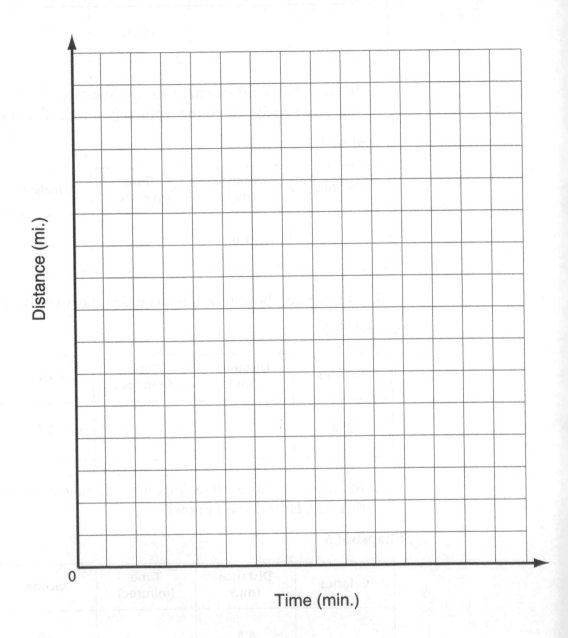

Test Practice

1. Celeste walks about $4\frac{1}{2}$ miles three times a week. If she walks an average of 3 miles per hour, how long does it take her to walk on any given day?

   (1) 45 minutes

   (2) 1.3 hours

   (3) 1.5 hours

   (4) 9 hours

   (5) 12 hours

2. Myra paced herself as she walked up and down the hall. The hall is 100 feet long. If it took her 40 seconds to walk the length of the hall, how many feet does she walk per second?

   (1) 0.4

   (2) 2.5

   (3) 4

   (4) 25

   (5) 60

3. Bronson's new car gets an average of 25 miles per gallon in highway driving and 12.5 miles per gallon in city driving. If he drives an average of 50 miles per week on the interstate and the same number of miles in the city, about how many gallons of gas does he use monthly?

   (1) 2

   (2) 4

   (3) 8

   (4) 16

   (5) 24

4. Justin ran 5,000 yards in 40 minutes. At this rate, about how many yards did he run per minute?

   (1) 0.008

   (2) 0.08

   (3) 8

   (4) 12.5

   (5) 125

5. Hue has to drive 345 miles by 8 p.m. If he drives about 55 miles per hour, what is the latest time (of the choices below) he can leave in order to arrive on time?

   (1) 9:00 a.m.

   (2) 11:30 a.m.

   (3) 12:00 noon

   (4) 1:30 p.m.

   (5) 2:00 p.m.

6. Ramona drove 90 miles in an hour and a half. At this rate, about how many miles will she have driven in 4 hours?

Playing with the Numbers

> **What happens when numbers are not so user-friendly?**

Two equal ratios form a proportion equation. A proportion equation always has four numbers, such as $\frac{2}{3} = \frac{12}{18}$. Sometimes you know three of the numbers and want to find the fourth. Whether you find that missing number in your head, with a calculator, or by using pencil and paper, it is important to be able to judge whether your answer is reasonable. When the numbers are unwieldy or difficult, it is helpful to round them first and then make an estimate. Estimating takes good number sense.

In this lesson, you will use another new tool for checking whether two ratios are equal. You will also play with more challenging numbers. You will be given three numbers and will find a fourth to complete the proportion. Estimation will come in handy.

Activity 1: Creating and Checking Equal Ratios

You know several ways to determine whether two ratios are equal. One way to check is with the **cross-product property**.

> The cross-product property: To check whether two ratios in fraction form are equal, multiply the top number of the first by the bottom number of the second. Then multiply the bottom number of the first by the top number of the second. If the products are equal, then the ratios are equal. If the products are not equal, then the ratios are not equal.
>
>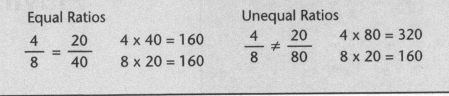
>
> | Equal Ratios | Unequal Ratios |
> |---|---|
> | $\dfrac{4}{8} = \dfrac{20}{40}$ $\quad$ 4 × 40 = 160 $\quad$ 8 × 20 = 160 | $\dfrac{4}{8} \neq \dfrac{20}{80}$ $\quad$ 4 × 80 = 320 $\quad$ 8 × 20 = 160 |

1. A True Proportion Equation

   a. Write a proportion equation that is true. You choose the numbers.

   b. Use the cross-product property to show that the ratios are equal.

   c. Use another way to show that the ratios are equal.

2. A False Proportion Equation

   a. Write a proportion equation that is false. You choose the numbers.

   b. Use the cross-product property to show that the ratios are not equal.

   c. Use another way to show that the ratios are not equal.

Activity 2: What Is the Fourth Number?

Card Set 1: Friendly Numbers

1. Place the four cards from Card Set 1 to form a proportion equation.

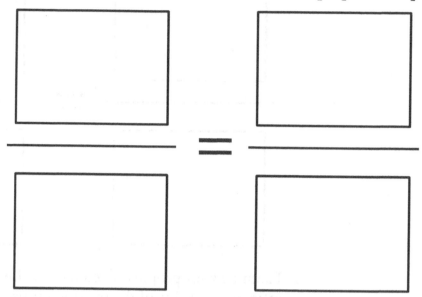

2. What value for the missing number would make the proportion true? Write it on the blank card.

3. How do you know you are right?

Card Set 2: More Challenging Numbers

1. Place the four cards from Card Set 2 to form a proportion equation.

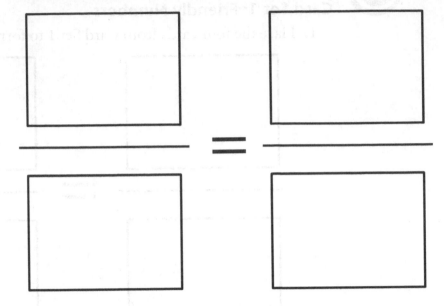

2. Examine your proportion equation. Use friendlier numbers to estimate the value of the missing number that would make the proportion equation true. Write your estimation next to the blank card.

3. Explain how you arrived at your estimation. What "friendly" numbers did you use?

4. What is a more exact value for the fourth number? Write the number you arrive at on the blank card.

Play with the numbers. Use a calculator to help with the unwieldy numbers.

5. Explain your method.

6. Is your final calculation close to what you expected?

Practice: Relating Multiplication and Division

Multiplication and division are closely related. For example, if 5 × 4 = 20, then 20 ÷ 5 = 4 and 20 ÷ 4 = 5. You can do that in your head, but when the numbers get larger, calculators help. Use a calculator to help complete each of the statements.

1. If 35 × _____ = 140, then 140 ÷ 35 = _____.

2. If 182 ÷ 13 = _____, then 13 × _____ = 182.

3. If 15 × 15 = _____, then _____ ÷ 15 = 15.

4. If 90 ÷ 12 = _____, then 12 × _____ = 90.

5. If 16 × _____ = 104, then 104 ÷ _____ = 16.

6. If _____ × 8.5 = 170, then 170 ÷ 8.5 = _____.

7. If 400 × _____ = 200, then 200 ÷ 400 = _____.

8. If 750 ÷ 25 = _____, then 25 × _____ = 750.

9. If 45 × _____ = 202.5, then 202.5 ÷ _____ = 45.

10. What do you notice about the relationship between multiplication and division?

Practice: What Is the Missing Number?

Figure out the missing number for each of the following proportions. Then check your answer by using another method.

1. $\dfrac{80}{\boxed{}} = \dfrac{20}{30}$

2. $\dfrac{\boxed{}}{50} = \dfrac{4}{25}$

3. $\dfrac{20}{35} = \dfrac{\boxed{}}{105}$

4. $\dfrac{20}{\boxed{}} = \dfrac{35}{105}$

5. $\dfrac{28}{60} = \dfrac{70}{\boxed{}}$

6. $\dfrac{5}{6} = \dfrac{\boxed{}}{240}$

Practice: Estimate and Solve

Figure out the missing number for each proportion by first estimating. Then find a more exact answer rounded to the nearest tenth.

	Estimate	**More Exact Number**

1. $\dfrac{26}{45} = \dfrac{\square}{92}$

2. $\dfrac{198}{103} = \dfrac{602}{\square}$

3. $\dfrac{\square}{81} = \dfrac{241}{163}$

4. $\dfrac{33}{\square} = \dfrac{133}{161}$

Practice: Patterns for Predicting

For each of the following problems, think about the relationship across the set of numbers (using the property of equal fractions), and predict what the missing number is. Then figure out the exact answer. Are your estimates close to the exact answers?

Example:

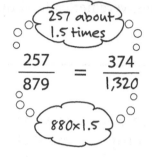

$$\frac{257}{879} = \frac{374}{1,320}$$

257 about 1.5 times

880×1.5

1.

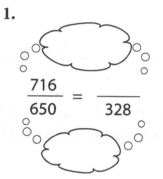

$$\frac{716}{650} = \frac{}{328}$$

2.

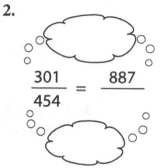

$$\frac{301}{454} = \frac{887}{}$$

3.

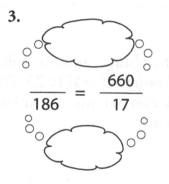

$$\frac{}{186} = \frac{660}{17}$$

4.

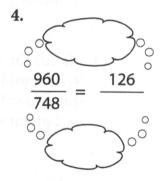

$$\frac{960}{748} = \frac{126}{}$$

Practice: Story Problems

Provide an estimate for each of the following problems. Then figure out the actual answer.

1. Carole has to figure out how much of the phone bill is her responsibility. The total minutes used were 542 for a cost of $37.94. If 281 minutes were her calls, about how much should she expect to pay?

   Estimate _____ Actual _____

2. On his last trip, Shawn drove 332 miles and used 12.8 gallons of gas. If his next trip requires that he travel 450 miles, about how many gallons should he expect to use?

   Estimate _____ Actual _____

3. Diego determined that he could walk 2,640 feet in about 10 minutes. At this rate, how many miles could he walk in an hour?

   Estimate _____ Actual _____

4. Katie's Kitchen buys hamburger meat in bulk. Last week the restaurant bought 75 pounds for $116.25. This week Katie's Kitchen will only order 60 pounds. About how much should the restaurant expect to pay?

   Estimate _____ Actual _____

Extension: Bringing Proportions to Life

Choose one of your true proportions from *Activity 2*, page 99, to create a real-life situation.

1. The proportion equation I chose:

2. Think about a situation those numbers could represent. Try to make it as realistic as possible. Add some labels (cost, kinds of items, miles, etc.) Write about the situation here.

Test Practice

1. Which of the following ratios is equal to the ratio $\frac{55}{65}$?

   (1) $\frac{13}{11}$

   (2) $\frac{25}{35}$

   (3) $\frac{27.5}{32.5}$

   (4) $\frac{110}{120}$

   (5) $\frac{130}{110}$

2. What would x have to represent in order to make the following proportion true?

   $$\frac{50}{40} = \frac{x}{220}$$

   (1) 5.5

   (2) 176

   (3) 200

   (4) 264

   (5) 275

3. What would x have to represent in order to make the following proportion true?

   $$\frac{x}{70} = \frac{30}{35}$$

   (1) 2.3

   (2) 15

   (3) 60

   (4) 210

   (5) 2,450

4. If 5.5 yards of cloth sells for $24.75, what would 2.5 yards cost?

   (1) $4.50

   (2) $9.00

   (3) $11.25

   (4) $13.75

   (5) $61.88

5. Which of the following proportions is true?

   (1) $\frac{10}{30} = \frac{60}{20}$

   (2) $\frac{10}{20} = \frac{30}{60}$

   (3) $\frac{10}{60} = \frac{20}{30}$

   (4) $\frac{60}{20} = \frac{10}{30}$

   (5) $\frac{10}{20} = \frac{60}{30}$

6. Find the missing number to make $\frac{75}{33} = \frac{\square}{11}$ a true proportion.

The Asian Tsunami

What did your money buy?

On December 26, 2004, an underwater earthquake caused **tsunamis** in Asia and Africa, affecting 11 countries. Other parts of the world reacted by sending donations to the stricken countries to help with food, water, medicine, and housing. What did the dollars the United States sent buy? How did U.S. dollars translate to other currencies such as Sri Lankan rupees, Indonesian rupiahs, and Thai bahts?

In this lesson, proportional thinking will come in handy when you **convert** from one country's **currency** to another's.

Activity 1: Tsunami Numbers in the News

What do you know about the Asian tsunami?

1. Use the following numbers to fill in the blanks in the story. Think about which numbers make sense.

500	2004	110,000
20	9.0	30,000
8,000	4.0	10,000

A tsunami triggered by a very large earthquake off the coast of the Indonesian island of Sumatra on December 26, _____, has left more than 150,000 people dead and millions homeless. Countries hit hardest by the disaster include Sri Lanka, Indonesia, India, Thailand, and the Maldives. Almost 75% of the deaths occurred in Indonesia, estimated at _____. Sir Lanka was second highest with about 20% of the estimated deaths, or _____ people lost that day. The rest of the deaths, approximately _____, occurred in the other nine countries affected by the tsunami.

The _____ foot wall of water, higher than a two-story building, swallowed entire villages. The tsunami waves were not only very high, they moved at a much faster speed than normal. These waves were comparable in size to those you see on some of the surfing movies; but those waves travel at 30 miles an hour, and the tsunami waves were moving more than fifteen times as fast at _____ miles an hour. The velocity of the force is what caused the destruction—a massive force that swept away everything in its path.

Earthquakes are measured on a Richter scale that has a range from 0 to 12. This was a destructive earthquake measuring _____ on the Richter scale, the fourth worst earthquake in recorded history. A 6.0 on the scale is a pretty bad earthquake. The Richter scale does not use direct proportion, so an 8.0 is not twice as strong as a _____ ; it is massively more. There are about _____ mini-quakes a day, which is almost three million per year.

(Story constructed from January 2005 news reports)

2. On the following map, locate and label the five countries mentioned in the story. Use an atlas. Make a guess as to where you think the epicenter of the earthquake was. Then look on the Internet to check your guess.

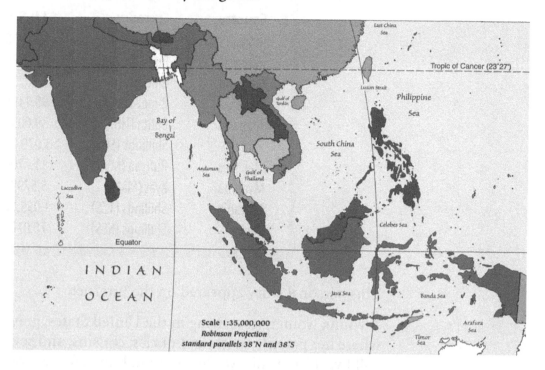

Activity 2: Tsunami Survivor Aid in Sri Lanka

Currency Conversion Table (January 2005)		
Country	Currency	$1 U.S. =
India	Rupees (INR)	59.8800
Indonesia	Rupiahs (INR)	9,297.99
Malaysia	Ringgits (MYR)	3.80776
Sri Lanka	Rupees (LKR)	98.1195
Thailand	Baht (THB)	39.0030
Somalia	Shillings (SOS)	3,079.00
Maldives	Rufiyaa (MVR)	11.7700
Myanmar	Kyat (MMK)	5.57991
Tanzania	Shillings (TZS)	1,055.00
Kenya	Shillings (KES)	78.0700

This personal story appeared on the Internet:

A young woman now living in the United States, grew up in Sri Lanka, where her parents, aunts and uncles, cousins, and several family friends still live. Her family was lucky enough to suffer no losses but heard about friends of friends who lost their lives. After the tsunami hit, the young woman made a list of what certain amounts of American money could buy in Sri Lanka.

For example, items that could be purchased for $1 each include an oil lamp, 10 candles, 10 boxes of matches, a pillow, a sleeping mat, a cooking pot, rice to feed a family of five for two days, half a pound of powdered milk, and two pounds of dried lentils. A small, three-room, cement-block home for a fisherman's family could be built for about $400; a fishing boat could be built for about $500; and an outboard motor could be purchased for about $1,000.

1. Rewrite the story in Sri Lankan rupees. Use rounded amounts.

   For example, items that could be purchased for about _____ rupees each include an oil lamp, 10 candles, 10 boxes of matches, a pillow, a sleeping mat, a cooking pot, rice to feed a family of five for two days, half a pound of powdered milk, and two pounds of dried lentils. A small, three-room, cement-block home for a fisherman's family could be built for about _____ rupees; a fishing boat could be built for about _____ rupees; and an outboard motor could be purchased for about _____ rupees.

The woman wrote another story a few weeks after the disaster hit:

"Today my mother's friend visited her. The friend had lived with her 26-year-old son and 25-year-old daughter in a small house in Palliyawatte, a village close to my home. Her house was entirely washed away while they were at church on the 26th.

"She is living with her brother at present, while her children are living with her sister. Her brother is allowing her to build a small house in his yard. The location of her former home is now covered with erosion protection rocks.

"Her requests are very simple. She said that she had received about 5,000 rupees in donations, which will be enough to erect some portion of a house. She said what she needed was cement for the floor and that her friends would help her with simple furnishings for the house.

"To give an idea about the cost of living here, 10,000 rupees is a living wage for a small family."

2. Convert the amounts of Sri Lankan rupees in the story to U.S. dollars.

   5000 rupees in donations is about _____.

   10,000 rupees is a living wage. That is about _____.

3. Explain the method you used to convert U.S. dollars to Sri Lankan rupees.

4. Explain the method you used to convert Sri Lankan rupees to U.S. dollars.

5. What is the U.S. dollar equivalent of one Sri Lankan rupee? How do you know?

Activity 3: What Your Money Buys

Oxfam America released this information on what various amounts of money could buy to help the tsunami relief effort:

$15 could pay for

- 70 packets of oral rehydration salts to treat diarrhea.

$20 could pay for

- Four long-lasting treated mosquito nets;

- Four Oxfam buckets—easily cleaned and with a tap for increased protection against contamination;

- One month's supply of soap for 120 people; or

- Basic hygiene kits for two families.

$50 could pay for

- A family food ration for one month; or

- A plastic latrine slab. When a flood or earthquake destroys a community's toilets, latrines are a quick way to prevent the spread of disease. They can be installed in a couple of hours.

$115 could pay for

- A 50-meter length of water distribution pipe.

$225 could pay for

- An emergency shelter kit for a family of eight containing plastic sheeting, pegs, and rope.

As concerned citizens of the world, your group wants to send a donation of $500 to one of the countries hit by the tsunami. Using the currency of the country, prepare a report, choosing the items for which you would wish your $500 to be used.

Report on Donating $500 (U.S.)

1. Country to receive the donation:

2. The currency of the country:

3. The exchange rate between the country and the United States:

4. Items we decided to donate and the estimated value of each in the country we chose:

Practice: Converting to U.S. Dollars

Find an international currency converter on the Internet. Use it to calculate today's exchange rate. Write the value in U.S. dollars of each amount of foreign currency shown below. Round the amount to the nearest cent.

1. 10 euros

2. 100,000 yen

3. 20 Canadian dollars

4. 50 Mexican pesos

5. 1 British pound

Practice: The Cost of a Big Mac®

McDonald's has restaurants in countries all over the world. Does a Big Mac cost the same everywhere?

1. Following are some prices (based on 2003 data) of Big Macs in different countries along with their exchange rates. Compare each country's price for a Big Mac with the price in the United States by completing the last column in the following table:

Country	Exchange Rate (= $1 U.S.)	Big Mac in Country's Currency	Big Mac in U.S. Dollars
United States	$1	$2.65	
Great Britain	0.53110	3.75 British pounds	
China	8.2665	9.95 yuan	
Indonesia	9,171	16,155 rupiah	
Mexico	11.217	22 pesos	
Thailand	38.562	55 bahts	

2. Which country seems to have the cheapest Big Mac? The most expensive?

Practice: At the Paris Airport

When you arrive at an airport in another country, one of the first things you might do is exchange your money. Use the exchange rate information to answer the following questions.

Conversions valid on February 1, 2005:

- 0.15245 euro = 0.19873 U.S. dollar (USD)

- 0.15245 U.S. dollar = 0.11695 euro (EUR)

- 1 euro = 135.060 Japanese yen

- 1 Japanese yen = 0.007404 euro

1. About how many euros could you convert from $250?

2. About how many euros could you convert from 300 Japanese yen?

Challenge:

3. From the information given, how can you figure out about how many Japanese yen there are in a dollar?

Extension: The Richter Scale and Earthquakes

To describe the strength of earthquakes, scientists use a scale of numbers called the Richter scale. The Richter scale grows by powers of 10. An increase of 1 point means the strength of a quake is 10 times greater than the previous level. This is how it works:

> An earthquake registering 2.0 on the Richter scale is 10 times stronger than a quake registering 1.0. A quake registering 3.0 is 10 x 10, or 100, times stronger than a quake registering 1.0. A 4.0 earthquake is 10 x 10 x 10, or 1,000, times greater than 1.0, and so on.

(Adapted from "Scholastic Math," December 1989)

The Asian tsunami was a massive earthquake at 9.0 on the Richter scale. It is the fourth worst earthquake in recorded history. A 6.0 on the scale is a pretty bad earthquake. The San Francisco earthquake in 1906 registered 8.0 on the Richter scale.

Ruins of City Hall, San Francisco, 1906. Library of Congress, Prints and Photographs Division, Detroit Publishing Company Collection.

1. Compare the strength of the earthquake of 2004 with that of the San Francisco earthquake of 1906.

Extension: Your Own Country

What experience do you have with currency conversion in another country? Write about it. What was the hardest thing to keep straight?

If you do not have a personal experience, interview someone who does.

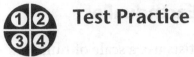

Test Practice

Note: Exchange rates are based on 2005 figures.

1. Hector noted that the exchange rate for Haitian gourdes was 1 U.S. dollar = 37.82100 Haitian gourde. At this rate, about how many gourdes could he get for $20?

   (1) 0.5

   (2) 50

   (3) 75

   (4) 380

   (5) 750

2. One euro is equivalent to about 1.6 Canadian dollars. At this rate, one Canadian dollar is worth about how many euros?

   (1) 0.16

   (2) 0.625

   (3) 6.25

   (4) 16

   (5) 62.5

3. One euro is equivalent to about 1.6 Canadian dollars. At this rate, 50 euros are worth about how many Canadian dollars?

   (1) 0.32

   (2) 32

   (3) 80

   (4) 160

   (5) 320

4. When Tai sent U.S. dollars to her relatives in Pakistan, they exchanged them for rupees. If 1 Pakistan rupee is worth 0.0168 U.S. dollars, about how many rupees did they get for $150?

   (1) 2.5

   (2) 11

   (3) 25

   (4) 112

   (5) 8,928

5. One month's Internet access costs about $40 in Australia (AUD). If $1 USD is about $1.61 AUD, about how much would this be in American dollars?

   (1) $3

   (2) $25

   (3) $32

   (4) $64

   (5) $800

6. One U.S. dollar is about 103.66 Japanese yen. At this rate, about how many dollars can you get for 5,000 yen? (Round to the nearest whole dollar.)

As If It Were 100

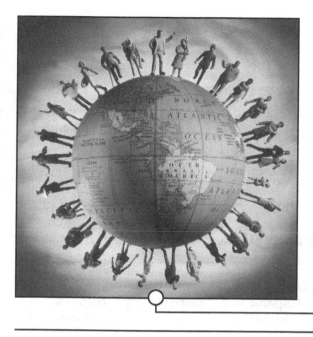

> *Why are percents so commonly used?*

Have you wondered why percents are used so often? One reason is that 100 is a very "friendly" number, one that makes it easy to compare amounts of different sizes. For example, when you are trying to figure out whether 9 women out of 10 people is a relatively more female population than 42,000 women out of 45,000 people, it is helpful to ask: What if the total population in each situation were 100?

In this lesson, you focus on percents as useful ratios. You will notice the power of percents for comparing data sets of different sizes. You will explore how to use 1,000 as a base of comparison as well.

Activity 1: How Does Your Program Compare?

The following statistics on the federal participation levels for adult education are based on the year 2003–2004, with a total of 2,677,028 participants in adult education classes.

Enrollment of Participants by Gender

Male	Female
1,223,883	1,453,145

Enrollment of Participants by Age

Age 16–18	Age 19–24	Age 25–44	Age 45–59	Age 60 and Over
372,584	677,499	1,200,608	328,558	97,779

Enrollment of Participants by Race/Ethnicity

Native American or Alaskan Native	Asian/Pacific Islander	Black, Not of Hispanic Origin	Hispanic	White, Not of Hispanic Origin
35,971	210,471	535,489	1,157,568	737,529

Source: U.S. Department of Education, Office of Vocational and Adult Education, Division of Adult Education and Literacy, March 2005

How does your program compare to adult education programs across America? Because your program has fewer learners than the number in the entire United States, it can seem difficult to see how your program compares nationally. Percents make the comparison easier because 100 is an easy number to use.

1. With your team, complete one section of the table (age or ethnicity) on the next page. Begin by entering the information about your own program. Then complete that section by filling in the national statistics.

Category	In Our Program		National Statistics	
	Number	% of total	Number	% of total
Total Enrolled				
Gender				
Men				
Women				
Age				
16–18				
19–24				
25–44				
45–59				
60+				
Ethnicity				
Native Amer.				
Asian/PI				
Black or African Amer.				
Hispanic				
White				

2. How does your program compare to the national statistics? Prepare a report that compares the national data for one of the categories (age or ethnicity) with your program's data. Include a circle graph in the report.

Activity 2: What If There Were 1,000?

If the world were a village of 1,000 people,

- 584 would be Asians;

- 123 would be Africans;

- 95 would be East and West Europeans;

- 84 would be Latin Americans;

- 55 would be Soviets (still including, for the moment, Lithuanians, Latvians, Estonians, etc.);

- 52 would be North Americans;

- 6 would be Australians and New Zealanders.

The people of the village would have considerable difficulty communicating.

- 165 people would speak Mandarin;

- 86 would speak English;

- 83 would speak Hindi/Urdu;

- 64 would speak Spanish;

- 58 would speak Russian; and

- 37 would speak Arabic.

That list accounts for the mother tongues of only half the villagers. The other half would speak (in descending order of frequency) Bengali, Portuguese, Indonesian, Japanese, German, French, and 200 other languages. In this global village, there would be

- 300 Christians (183 Catholics, 84 Protestants, 33 Orthodox),

- 175 Moslems,

- 128 Hindus,

- 55 Buddhists,

- 47 Animists, and

- 210 people of all other religions (including atheists).

Source: *State of the Village Report, The Donella Meadows Archive: Voice of a Global Citizen*, Sustainability Institute http://www.sustainer.org

1. What strikes you about the description of a world of 1,000 people?

2. Decide upon the category that most interests you. Then use what you know about ratios to figure out how many people are actually in that group, as well as the percent of the world population that it represents. Look up the present world population on the Internet.

   Write a story about the category you chose. Be sure to include comparisons using actual numbers and percents.

3. Show how you know that some of your ratios are equal.

Practice: Percent Estimates

Round each of the sets of numbers to make them friendlier and then estimate the percent.

Example: 1,234 out of 12,400

$$1,200 \text{ out of } 12,000 = 10\%$$

1. 123 out of 251

   _____

2. 435 out of 1,246

   _____

3. 1,539 out of 2,031

   _____

4. 988 out of 1,001

   _____

5. 383 out of 1,203

   _____

6. 1,212 out of 2,431

   _____

7. 48 out of 478

   _____

8. 512 out of 2,531

   _____

9. 696 out of 7,032

   _____

Practice: Percent Dilemmas

Choose one of the following percent dilemmas and describe what the teacher is trying to explain. How would you show the student what is happening?

Dilemma 1

Teacher: "There are 6 people in this class; 3 are native born. So 50% of the class was born in America."

Student: "I can see the '3' out of the bunch of us, but where do you get the number 50?"

Teacher: "Well, we don't really have 50 people. It's just *as if* we had 100 people."

Dilemma 2

Student: "I can't believe it. You said that the number of men in adult education is about the same as the number of women. That's not what I get. There are about 200,000 more women than men according to the statistics I see."

Teacher: "But when you look at the percents, you can see how relatively close the number of men and women are."

Dilemma 3

Student: "I can see that the ratio of teachers to students is 1:4, but I can't see how you can get 80% from that. It seems like it should be 75% instead."

Teacher: "In this case, the 80% represents the portion of students out of the total number of people."

Practice: Free Throws—Who Is the Best?

Three men were practicing free throws.

- Tom hit 16 of 25 of his shots.

- Jerome got 18 of 40 of his shots.

- Kayem made 15 out of 20 of his shots.

1. Who is the best free-throw shooter? Jerome says he is because he sank the most shots. Tom says he is because he got more than Kayem but missed less than Jerome. What do you say?

2. Explain how you arrived at your decision.

Practice: The Savings Contest

Tai, Shaina, and Margot decided to have a contest to see who could save the most money over a six-month period. Each chose a different savings goal based on her personal situation.

At the end of the six months, they compared how close they came to reaching their goals.

- Tai's goal was to save $800. She saved $360.

- Shaina's goal was to save $600. She saved $300.

- Margot's goal was to save $1,500. She saved $1,000.

1. Margot says she won the contest because she came closest to her goal. Do you agree?

2. Explain your decision and show how you compared the three goals.

Practice: Pet Store Survey

The Pick-a-Pet Store at the mall took a survey of adults and their pets. The results of the survey are shown in the following chart:

	Male	Female	Age 21–35	Age 36+
Dog Owners	48	22	12	58
Cat owners	24	56	26	54
Bird Owners	8	12	10	10
Total	80	90	48	122

Write a brief report based on the results of the survey. Include at least two statements that use percents and one that includes a ratio in another form.

Practice: Comparing Your Program to the Community Adult Education Program

Look at the following chart and think about how the adult education students in your program compare with those in the Community Adult Education Program. Use percents to make your comparisons.

Community Adult Education Program

	Male	Female	16–18	19–24	25–44	45–59	60+
Asian	9	14	3	2	10	8	0
Black or African Amer.	12	14	3	3	15	3	2
Hispanic	10	13	5	10	2	6	0
White	5	8	3	3	5	0	2
Total							

Extension: Parts per Million

Percents are based on 100, but businesses sometimes want to make comparisons using larger numbers. Often they focus on a million, rather than just one hundred.

1. Consider a business that says that it is producing quality products 99% of the time. Do you think that a company that makes good products 99% of the time is effective?

2. Fill in the table to show how many products would be good and how many would be rejects.

Total Products	Number of Good Products	Number of Rejects	Percent of Good Projects
a. 100			99%
b. 1,000			99%
c. 10,000			99%
d. 100,000			99%
e. 1,000,000			99%

3. What happened as the number of products increased?

4. Does 99% quality appear as good when you focused on 1,000,000 products as when you were considering only 100 products? Why?

Test Practice

1. At the County Education Program, there are a total of 89 adults enrolled. Twenty-three of them are in the GED classes, and the rest are in the English language classes. About what percent of the students in the program are in English language classes?

   (1) 4%

   (2) 35%

   (3) 59%

   (4) 66%

   (5) 74%

2. In the island community of Nomeport, there are 3,453 Native Americans, 1,487 Whites, and 2,560 Asians. About what percent of the island is Native American?

   (1) 12%

   (2) 22%

   (3) 46%

   (4) 54%

   (5) 85%

3. In the quality control department, there are 15 Hispanics, 12 Blacks or African Americans, 13 Whites, and 8 Asians. About what percent of the quality control department is *not* Asian?

   (1) 8%

   (2) 17%

   (3) 20%

   (4) 40%

   (5) 83%

4. At Basic Building, there are 12 female and 46 male employees. About what percent of the employees are female?

   (1) 4%

   (2) 21%

   (3) 26%

   (4) 58%

   (5) 79%

5. According to a survey of 100 women and 95 men, 74 women and 61 men preferred cats to dogs. Based on the survey what percent of males preferred cats to dogs?

   (1) 31%

   (2) 64%

   (3) 69%

   (4) 74%

   (5) 82%

6. According to a survey of 100 women and 95 men, 74 women and 61 men preferred cats to dogs. Based on the survey, about what percent (rounded to the nearest percent) of people polled preferred cats to dogs?

Closing the Unit: Reasoning with Ratios

> *Where do you find rates and ratios outside the math classroom?*

Rates and ratios can be found all around us. What are some situations in which you encounter them in your own life?

In this session, you will carry out a final project by going outside the classroom to search for rates and ratios. You will draw upon your own personal experience and create a report to share with the rest of the class. Then you will complete some tasks that continue to call upon your ability to reason with ratios.

Activity 1: Rates and Ratios In Your Life

For this final project, you are going to search for evidence of rates and ratios in your own life and then prepare a report on what you find. You might investigate at home, at work, or in your community. Here are some suggestions for places to look for rates and ratios, but do not limit yourself to these. Be creative and have fun as you search. Here are some places you might go to conduct your investigation:

- A shopping mall

- A gas station

- A doctor's office or your medicine cabinet

- The newspaper

- A bank

- Your car

- A gym

- The mail, including utility and telephone bills

- A place that sells lottery tickets

In your report, include the following:

1. Where and what you investigated.

2. Several rates and ratios you discovered. Include actual items you found, or take a photo or draw a sketch.

3. Choose one of the ratios to focus on. Predict for at least two other amounts based on the ratio that you chose. Draw out your prediction and make a set of equal fractions to show how you made your prediction. Also show the relationship with a graph.

4. Are there any situations you investigated that you found were not proportional? If so, describe them; if not, think of a situation that would not be proportional.

5. Using the information you found, write a math word problem for someone else to solve.

6. Looking closely at your exploration, what did you learn?

Activity 2: Final Assessment

Complete the tasks on the *Final Assessment* your teacher will give you.

Activity 3: Mind Map

Make a Mind Map listing everything that comes to mind when you think of *rate*, *ratio*, and *proportion*. When you finish, compare your first Mind Map, page 4, with this final one.

What did you notice when you made the comparison?

VOCABULARY

Lesson	Terms, Symbols, Concepts	Definitions and Examples
Opening the Unit	comparison	
	per	
	predict	
1	fraction form	
	property of equal fractions (or ratios)	
	proportion	
	rate	
	ratio	
2	sample	
3	concentrate	
	mixture	
4	percent	

VOCABULARY (continued)

LESSON	TERMS, SYMBOLS, CONCEPTS	DEFINITIONS AND EXAMPLES
6	area	
	dimensions	
	perimeter	
	surface area	
7	speed	
	unit rate	
8	cross-product property	
9	convert	
	currency	
	tsunami	

REFLECTIONS

OPENING THE UNIT: Comparing and Predicting

If someone told you, "Make sure you keep things in proportion," what would they be talking about?

_____

_____

_____

_____

_____

_____

_____

LESSON 1: A Close Look at Supermarket Ads

In your own words, explain how to keep ratios the same. Then give some examples of situations when ratios do not stay the same.

_____

_____

_____

_____

_____

_____

_____

LESSON 2: It's a Lot of Work!

Explain how ratios are useful in making predictions from samples. What are some ways that you make decisions in your life based on samples?

_____

_____

_____

_____

_____

_____

_____

LESSON 3: Tasty Ratios

What did you learn today that helped you deepen your understanding of ratios?

_____

_____

_____

_____

_____

_____

_____

LESSON 4: Another Way to Say It

Think of an example of an advertisement that you have heard on the radio or seen on a billboard or in a magazine. What are some other ways that the advertisers could have presented the information?

_____

_____

_____

_____

_____

_____

_____

_____

LESSON 5: Mona Lisa, Is That You?

What are some ways that you have seen ratios used in visual representations? How did the ratios influence the visual representation?

_____

_____

_____

_____

_____

_____

LESSON 6: Redesigning Your Calculator

What happens to the area of an object when you double the dimensions? Explain why this is so.

_____

_____

_____

_____

_____

_____

_____

LESSON 7: Comparing Walks

Explain how to find a unit rate. What other examples can you think of in which rates are described in terms of a unit rate?

_____

_____

_____

_____

_____

_____

_____

LESSON 8: Playing with the Numbers

Explain in your own words how to use the cross-product property to check whether two ratios are equal. Also, explain how to use estimation when working with messy numbers.

_____

_____

_____

_____

_____

_____

_____

LESSON 9: The Asian Tsunami

Explain in your own words how to convert currency. Also, explain why using estimation is important when working with messy numbers.

_____

_____

_____

_____

_____

_____

_____

LESSON 10: As If It Were 100

How are percents useful in making comparisons? Give an example or two to make your point clear.

_____

_____

_____

_____

_____

_____

_____

CLOSING THE UNIT: Reasoning with Ratios

What are the most important ideas and skills you have learned and will remember about rates, ratios, and proportions?

_____

_____

_____

_____

_____

_____

_____